Douglas G. Brookins

Eh-pH Diagrams for Geochemistry

With 98 Figures and 61 Tables

Springer-Verlag
Berlin Heidelberg New York
London Paris Tokyo

Dr. Douglas G. Brookins
Department of Geology
University of New Mexico
Albuquerque, NM 87131, USA

ISBN 3-540-18485-6 Springer-Verlag Berlin Heidelberg New York
ISBN 0-387-18485-6 Springer-Verlag New York Berlin Heidelberg

Library of Congress Cataloging-in-Publication Data. Brookins, Douglas G. Eh-pH diagrams for geochemistry.
Includes Index. 1. Geochemistry, Analytic. 2. Hydrogen-ion concentration. I. Title.
QE516.3.B76 1987 551.9 87-36931.

This work is subject to copyright. All rights are reserved, whether the whole or part of the material is concerned, specifically the rights of translation, reprinting, re-use of illustrations, recitation, broadcasting, reproduction on microfilms or in other ways, and storage in data banks. Duplication of this publication or parts thereof is only permitted under the provisions of the German Copyright Law of September 9, 1965, in its version of June 24, 1985, and a copyright fee must always be paid. Violations fall under the prosecution act of the German Copyright Law.

© Springer-Verlag Berlin Heidelberg 1988
Printed in the United States of America

The use of registered names, trademarks, etc. in this publication does not imply, even in the absence of a specific statement, that such names are exempt from the relevant protective laws and regulations and therefore free for general use.

Typesetting: K+V Fotosatz GmbH, Beerfelden
2131/3130-543210

Preface

Eh-pH diagrams have been in moderate use since the late 1950s in the fields of geochemistry, metallurgy, and chemical engineering; more recently, they have been popular in addressing aspects of radioactive and other hazardous waste disposal problems. Yet in many instances these diagrams are outdated and incomplete. In other cases the behaviour of the chemical elements in pure water is considered, yet no attention is given to the possible presence of dissolved sulfur and/or carbon dioxide species. In still other cases the data used are old, and lead to erroneous diagrams.

In this book I have attempted to prepare generic Eh-pH diagrams for the chemical elements and their species in the presence of water and with, where relevant, the presence of dissolved sulfur and carbon dioxide. Some 75 elements are covered here; only the alkali elements, the noble gases, and a few radioactive elements for which there are insufficient thermodynamic data, have been omitted. I include here those elements with more than one valance in natural waters, and those monovalent elements are sensitive to sulfide-sulfate redox conditions. In addition, I have plotted the stability fields of solids and aqueous species of pH-dependent only elements in Eh-pH space so that the reader may compare these conditions for a great variety of species stable under the Eh-pH limits of water.

For many years I have been bothered by the fact that Eh-pH diagrams, and the data used to construct them, are scattered throughout a large volume of literature. This book attempts to gather all the good to very good thermodynamic data useful for the 75 elements considered, and to construct the appropriate Eh-pH diagrams. A brief text accompanies each element that is designed to explain the diagram for that element, but is not in any sense intended to be a comprehensive discussion of the geochemistry of the elements. In many cases I have added a brief text on the importance of various elements to problems of hazardous waste disposal.

The diagrams have been constructed for 25 °C and 1 bar pressure conditions. A forthcoming project will address Eh-pH conditions at higher temperatures and pressures, as I do not feel the thermodynamic data base is adequate to do this yet for many elements. The thermodynamic data base used for this book is, admittedly, uneven, in that some older and less precise data are used as well as recent and high quality data. This, of course, is necessary to allow the calculation of the diagrams, but will result in revisions of many of the diagrams as never and more refined data become available.

The goal of this book is also to allow the reader an opportunity to inspect the role of redox reactions in nature, and to compare various mineral assemblages and/or aqueous species in Eh-pH space. This will allow a greater appreciation of the utilization of these diagrams in geochemistry and other fields, especially since it is all packaged in one place.

Albuquerque, February 1988   *Douglas G. Brookins*

Acknowledgments

Modern Eh-pH diagrams are built on the foundation provided by M. Pourbaix (Pourbaix 1966) and R.M. Garrels (Garrels 1959; Garrels and Christ 1965), and I respectfully and gratefully acknowledge their monumental efforts on this subject. Partial financial support was made available for some of this project by the Oak Ridge National Laboratory, where I spent part of my 1986–1987 sabbatical leave, and from the University of New Mexico Research Allocations Committee. Mr. D. Lopez drafted many of the figures. This project could not have been completed without the painstaking yet expeditious efforts of Ms. Judith Binder of the University of New Mexico, who prepared the final manuscript and co-checked proofs. Her efforts are especially appreciated.

Contents

Introduction 1
    Previous Studies 2
    Eh Versus pE 2
    Eh-pH Diagrams: Background 3
    Types of Boundaries in Eh-pH Space 4
    Thermodynamic Data and Uncertainties 7
    Use of Eh-pH Diagrams at Other Temperatures and Pressures 8
    Stability Limits for Water 8
    Natural Waters 9
    Presentation of the Elements and Elements Covered 12

Iodine and Other Halides	14	Iron	73
Sulfur	16	Palladium	82
Selenium	18	Rhodium	84
Tellurium	20	Ruthenium	86
Polonium	22	Platinum	88
Nitrogen	24	Iridium	90
Phosphorus	26	Osmium	92
Arsenic	28	Manganese	94
Antimony	30	Technetium	97
Bismuth	32	Rhenium	100
Carbon	34	Chromium	102
Silicon	36	Molybdenum	104
Germanium	38	Tungsten	106
Tin	40	Vanadium	108
Lead	42	Niobium	110
Boron	44	Tantalum	112
Aluminum	46	Titanium	114
Gallium	48	Zirconium	116
Indium	50	Hafnium	118
Thallium	52	Scandium	120
Zinc	54	Yttrium and the Rare Earth Elements (REE)	122
Cadmium	56	Americium	141
Mercury	58	Plutonium	144
Copper	60	Neptunium	146
Silver	64	Uranium	151
Gold	66	Thorium	158
Nickel	68	Beryllium	160
Cobalt	71		

Magnesium .... 162 Barium 168
Calcium .... 164 Radium 170
Strontium 166

References 172

Subject Index 175

Introduction

The use of Eh-pH diagrams in geochemistry is widespread, and their application to many aspects of mineral paragenesis well documented. These diagrams have been proven useful for metallurgical purposes, for predicting species in solutions, for alteration products of ores and in hydrothermal systems, and many others.

One problem with these diagrams is that, with one exception (Pourbaix 1966), no attempt has been made to compile the necessary thermodynamic data in order to treat as many elements as possible in one source. In this book I have attempted to do this. I have chosen data from the open literature that is considered to be highly reliable in many if not most cases (the exceptions, which at times must be included to allow certain calculations to be made), and the diagrams have been constructed for 25 °C, 1 bar pressure conditions. My attempt here is to provide generic diagrams for the elements where redox reactions are important, or where plotting dissolution, etc. of species in Eh-pH space is useful.

Eh-pH diagrams are, however, limited in their applicability to many geochemical processes. Complex solid solutions are not, in most cases, amenable to treatment due to lack of good thermodynamic data. Kinetic factors can cause metastable assemblages that are not consistent with Eh-pH predicted stabilities. In many cases aqueous species of various elements are not well characterized and, in fact, may not even have reliably identified; hence, most diagrams in this book may be subject to revision as new and better identification and data become available.

Nevertheless, it is possible to calculate a working diagram for most of the naturally occurring elements and even some totally radioactive elements as well. I have done this in this book and I hope my exercise is justified. Accompanying each diagram is a table with good to very good thermodynamic data compiled from the recent literature, and a brief text describing the characteristics of the diagram. It is not the purpose of this book to provide an in-depth discussion of the geochemistry of the various elements, and the reader is referred to the general geochemical literature for such detail.

Finally, I have attempted to cite relevant Eh-pH sources wherever possible, but it is certain that I have either intentionally or unintentionally omitted possibly "favored" diagrams of some readers. Hopefully, this will not cause undue problems to the readers.

An outgrowth of this effort will be a study of Eh-pH diagrams at temperatures and pressures different than 25 °C, 1 bar pressure, as data and ways to calculate data are now becoming available for such a project.

Previous Studies

There have been many, many Eh-pH diagrams (including pE-pH diagrams) published in the last 3 decades. The most comprehensive of these is the work of Pourbaix (1966) who compiled Eh-pH diagrams for the elements in his classic study of metal corrosion. Pourbaix (1966), however, was primarily concerned with reactions of metals and compounds in the presence of water only, and thus did not attempt to consider the effect of total dissolved carbonate, sulfur, or other species. In the United States of America, the most vigorous work on Eh-pH diagrams has been championed by R.M. Garrels (Garrels 1959; Garrels and Christ 1965). He has also discussed the derivation of the diagrams and their application to a wide variety of geologic problems in these works. In addition, good discussions of Eh-pH diagrams and their uses have been given by Krauskopf (1979), Berner (1971), Stumm and Morgan (1981), and others. The writer has published a number of diagrams over the last 10 years (Brookins 1978a–c, 1979c, 1983a, 1984, 1986a, b, 1987a, b), many of which were derived for application to aspects of radioactive waste disposal. Numerous diagrams have been published in the open literature, many of which will be cited as specific elements are discussed in the following pages. Because of the large volume of geochemical literature, it is certain that I have missed some diagrams, but hopefully none that would radically change the overall diagrams presented herein.

Eh Versus pE

The choice of Eh instead of pE (i.e., the negative log of the electron) is arbitrary. In choose Eh in this book because most geochemists use Eh-pH diagrams rather than pE-pH diagrams, and, further, Eh is a measurable quantity whereas pE is not. The relationship between Eh and pE is given by:

$$pE = \frac{F}{2.3\,RT} Eh , \qquad (1)$$

where F is the Faraday constant, R the gas constant, and T the absolute temperature.

Measurement of Eh is beyond the scope of this book. Good treatments of the problems associated with Eh measurements are given by Garrels and Christ (1965), Krauskopf (1979), Stumm and Morgan (1981), and others. Not only are there problems with these measurements, there are also many situations in nature where various half-cell couples in waters are not in equilibrium. This has been pointed out by Lindberg and Runnels (1984), who note that sulfur couples do not agree with carbon couples and other natural couples, hence local equilibrium (i.e., overall large-scale disequilibrium) controls many reactions in nature. While this is now a recognized fact, it is also a recognized fact that the very repetitive nature of mineral occurrences on the earth's surface and in the near-surface are readily explainable in many instances by the use of Eh-pH diagrams. Hence, while factors such as local equilibrium and kinetics make across-the-board use of Eh-pH diagrams unwise, their use is well demonstrated for explaining mineral parageneses, probable reactions in natural waters, and related areas. Hostettler (1984) has also discussed problems of Eh and related topics in natural waters.

Eh-pH Diagrams: Background

Eh-pH diagrams have been proven to be very useful in geochemistry. Some previous work will be summarized later in this chapter. For now, the background for the diagrams will be given.

Many reactions in nature involve oxidation and reduction. One has a choice of attempting to write complete reactions where the total amount of oxidation equals the total amount of reduction, or instead working with half-cell reactions for particular species of interest. The latter approach has the advantage that a large number of half-cells are known, or can be determined quickly, whereas the overall oxidation-reduction processes in the whole system are often too complex to treat quantitatively.

For a typical oxidation-reduction (from here on called redox) half-cell, one writes:

reduced species = oxidized species + (n) electrons ,

where n is the number of electrons. All such reactions have a certain standard potential, E^0. This parameter is related to the free energy of the half-cell reaction by:

$$\Delta G_f^0 = nFE^0 , \tag{2}$$

where ΔG_f^0 is the Gibbs free energy of the reaction, F is the Faraday constant, and E^0 is the standard potential. Consequently, we can rewrite the expression for E^0 by:

$$E^0 = \frac{-RT \ln K_e}{nF} , \tag{3}$$

since $\Delta G = -RT \ln K_e$, where R is the gas constant, T is the absolute temperature, and K_e is the equilibrium constant for a particular reaction. Thus, for a reaction of:

$$wA + xB = yC + zD + ne^- , \tag{4}$$

then

$$K_e = \frac{(C)^y (D)^z (e)^n}{(A)^w (B)^x} , \tag{5}$$

where the terms in parentheses are activities. For unit activities, one obtains the simple expression $\Delta G_f^0 = nE^0F$ from which E^0 can be calculated directly from the knowledge of the free energies of formation of species A, B, C, D and their operators w, x, y, z. The activity of the electron is always taken as unity and its ΔG_f^0 is assumed zero.

The potential Eh at conditions of other than unit activities of the species involved can be determined from the Nernst equation, viz.:

$$Eh = E^0 + \frac{RT}{nF} \ln (K_e) \ . \tag{6}$$

The K_e term is calculated from the activities of A, B, C, D to their w, x, y, z powers. Since many of the terms in the Nernst equation are constants (R, T, F) we can rewrite Eq. (6) as

$$Eh = E^0 + \frac{0.059}{n} \log \frac{(C)^y (D)^z}{(A)^w (B)^x} \ . \tag{7}$$

The 0.059 figure results from using $R = 0.001987$ kcal/deg, $T = 298.15°$, $F = 23.06$ kcal/volt-gram equivalent, and 2.303 to convert the natural log to log base 10.

Now assume that species D is actually H^+. In this case it is necessary to expand the $\frac{0.059}{n} \log (K_e)$ term into a term containing H and a term for the other species, viz.:

$$Eh = E^0 + \frac{0.059}{n} \log \frac{(C)^y}{(A)^w (B)^x} + \frac{0.059}{n} \log (H^+)^z$$

and

$$Eh = E^0 + \frac{0.059}{n} \log \frac{(C)^y}{(A)^w (B)^x} - \left(\frac{z}{n}\right)(0.059)\text{pH} \ . \tag{8}$$

In a subsequent section, examples of slopes or H^+ as a reactant as well as a product are given.

Types of Boundaries in Eh-pH Space

There are actually relatively few types of boundaries in Eh-pH space. Most are straight lines that are functions of pH alone, Eh alone, or positively or negatively sloped Eh- and pH-dependent lines. Curved boundaries appear when there is a change in the concentration of one species relative to another, i.e., HCO_3^{1-} to CO_3^{2-}, as an oxidation-reduction reaction takes place. Below are listed some of the more common types of boundaries.

1. pH-dependent only

   $$M_2O_3 + 6H^+ \rightleftarrows 2M^{3+} + 3H_2O .$$

   Solution: from $\Delta G_R^0 = -1.364 \log K$, and assuming an activity for M^{3+} solve for pH

   $$(H)^6 = (M^{3+})^2 K^{-1}; \quad pH = -\log \frac{[(M^{3+})^2(K^{-1})]}{6} .$$

2. Eh-dependent only: metal to ion

   $$M \rightleftarrows M^+ + e^{1-} .$$

   Solution: from $nE^0 F = \Delta G_R^0$, solve for E^0; then use Nernst expression for Eh:

   $$Eh = E^0 + \frac{0.059}{n} \{\log (M^+)\} .$$

3. Eh-dependent only: ion to ion

   $$M^{2+} \rightleftarrows M^{3+} + e^{1-} .$$

   Solution: from $nE^0 F = \Delta G_R^0$, solve for E^0; then use Nernst expression for Eh:

   $$Eh = E^0 + \frac{0.059}{n} \log \frac{(M^{3+})}{(M^{2+})} = E^0$$

   and $Eh = E^0$ since the activities of M^{2+} and M^{3+} are equal ($\log 1 = 0$) .

4. Eh-pH dependent: activities of redox species equal

   $$M^{3+} + 2H_2O \rightleftarrows MO_2^{1+} + 4H^{1+} + 2e^{1-}$$

   $$Eh = E^0 + \frac{0.059}{n} \log \frac{(MO^{1+})}{(M^{3+})} + \frac{0.059}{n} \log (H^{1+})^4$$

   $$Eh = E^0 - 0.118 \, pH .$$

5. Eh-pH dependent: solid-ion redox reaction

   $$MO + 2H^{1+} \rightleftarrows M^{3+} + H_2O + e^-$$

   $$Eh = E^0 + \frac{0.059}{n} \log (M^{3+}) + \frac{0.059}{n} \log \left[\frac{1}{(H^{1+})^2} \right]$$

   $$Eh = \left\{ E^0 + \frac{0.059}{1} \log (M^{3+}) \right\} + 0.0295 \, pH .$$

(Note: slope of this boundary is positive since H^{1+} ions appear with reactants.)

6. Eh-pH dependent: solid-ion redox reaction

$$MO + H_2O \rightleftarrows MO_2^{1-} + 2H^{1+} + e^{1-}$$

$$Eh = \left\{ E^0 + \frac{0.059}{1} \log (MO_2^{1-}) \right\} - 0.118 \, pH.$$

(Note: slope of this boundary is negative since H^{1+} ions appear with products.)

7. Eh-pH dependent: ion-solid redox reaction

$$2M^{2+} + 3H_2O \rightleftarrows M_2O_3 + 6H^{1+} + 2e^{1-}$$

$$Eh = E^0 + \frac{0.059}{n} \log \frac{1}{(M^{2+})^2} \frac{0.059}{n} \log (H^{1+})^6$$

$$Eh = \left\{ E^0 + \frac{0.059}{2} \log \frac{1}{(M^{2+})^2} \right\} - 0.177 \, pH.$$

8. Eh-pH dependent: more than one element involved in redox reactions.

a) Oxidation of metal sulfide:

$$2MS + 11H_2O \rightleftarrows M_2O_3 + 2SO_4^{2-} + 22H^{1+} + 18e^{1-}$$

$$Eh = E^0 + \frac{0.059}{n} \log (SO_4^{2-})^2 + \frac{0.059}{n} \log (H^{1+})^{22}$$

$$Eh = \left\{ E^0 + \frac{0.059}{18} \log (SO_4^{2-})^2 \right\} - 0.072 \, pH.$$

b) Oxidation of M ion and S(-II) sulfur:

$$2M^{2+} + 2H_2S + 11H_2O \rightleftarrows M_2O_3 + 2SO_4^{2-} + 26H^{1+} + 18e^{-1}$$

$$Eh = E^0 + \frac{0.059}{n} \log \left[\frac{(SO^{2-})^2}{(M^{2+})^2} \right] + \frac{0.059}{n} \log (H^{1+})^{26}$$

$$Eh = \left\{ E^0 + \frac{0.059}{18} \log \left[\frac{SO^{2-})^2}{(M^{2+})^2} \right] \right\} - 0.085 \, pH.$$

c) S oxidized, M reduced

$$MS + 4H_2O \rightleftarrows M^0 + SO_4^{2-} + 8H^{1+} + 6e^{1-}$$

$$Eh = E^0 + \frac{0.059}{n} \log (SO_4^{2-}) + \frac{0.059}{n} \log (H^{1+})^8$$

$$Eh = \left\{ E^0 + \frac{0.059}{6} \log (SO_4^{2-}) \right\} - 0.079 \, pH.$$

9. Eh-pH dependent: curved boundaries:

a) M ion − M carbonate

$$M^{1+} + \begin{bmatrix} H_2CO_3 \\ HCO_3^{1-} \\ CO_3^{2-} \end{bmatrix}^1 \rightleftarrows MCO_3 + \begin{bmatrix} 2H^{1+} \\ 1H^{1+} \\ OH^{1+} \end{bmatrix} + e^{-1}$$

(Note: at pH = 6.4 use $\frac{1}{2}H_2CO_3 + \frac{1}{2}HCO_3^{1-}$ and
at pH = 10.3 use $\frac{1}{2}HCO_3^{1-} + \frac{1}{2}CO_3^{2-}$)

$$Eh = E^0 + \frac{0.059}{n} \log \frac{1}{\Sigma CO_2} + \frac{0.059}{n} \log (H^{1+})^x ,$$

where ΣCO_2 = dominant dissolved carbonate species and x has values from 0 to 2, the slope will change from −0.118 pH (H_2CO_3 dominant) to 0 (CO_3^{2-} dominant).
(Note: this curve is concave).

b) Metal carbonate: metal oxide

$$2MCO_3 + 3H_2O \rightleftarrows M_2O_3 + 2\begin{bmatrix} H_2CO_3 \\ HCO_3^- \\ CO_3^{2-} \end{bmatrix}^1 + \begin{bmatrix} 2H^+ \\ 4H^+ \\ 6H^+ \end{bmatrix} + 2e^- ,$$

where slopes are:

0.059 pH, 0.118 pH, 0.177 pH, respectively.

(Note: this curve will be convex).

c) Metal-metal sulfide

$$M + \begin{Bmatrix} H_2S \\ HS^- \end{Bmatrix} \rightleftarrows MS + \begin{Bmatrix} 2H^+ \\ H^+ \end{Bmatrix} + 2e^- ,$$

where slopes are: −0.059 pH and −0.0295 pH, respectively.

Thermodynamic Data and Uncertainties

The thermodynamic data used in the preparation of the Eh-pH diagrams in this book are from many sources. Where possible, I have tried to use well-established sources, such as the National Bureau of Standards compilations (Wagman et al. 1982), the US Geological Survey (Robie et al. 1978), and others. For those species of interest to the field of radioactive waste disposal, the recent and on-going compilation of the OECD (Organization for Economic Cooperation and Development 1985) is especially valuable.

---

[1] The exact budget of different carbonic acid species must be calculated if (pH) to rigorously solve this equation.

All data used in this book are reported in units of kilocalories per gram formula weight (gfw) for a particular species. This is done primarily to make the diagrams consistent with the overwhelming majority of Eh-pH diagrams based on ΔG_f^0 data, only these data have been tabulated herein. The reader is referred to the original reference for tabulation of ΔH_f^0, S_f^0, and C_p data.

The uncertainties in the thermodynamic data are not treated in depth here. For the National Bureau of Standards data (Wagman et al. 1982), the uncertainty is between 8 and 80 units of the right-hand-most digit. For most species, this introduces only a small uncertainty in the calculations involving the particular data. For other species, I have referred the reader to the reference material.

Since the Eh-pH diagrams are generic, and since I have used data available for stoichiometric species, the uncertainties for most species have but a very small effect on the resultant calculations. This effect is also small then on the boundaries in Eh-pH space.

Use of the Eh-pH Diagrams at Other Temperatures and Pressures

Following geochemical convention, all the Eh-pH diagrams in this book are calculated for 25 °C (298.15 K) and 1 bar (10^5 Pa) pressure. To calculate the Eh-pH diagrams for any other temperature, but still at 1 bar, pressure is straightforward. In some instances (Barner and Scheuerman 1978; Robie et al. 1978; others) ΔG_f^0 data are reported for temperatures other than 25 °C. In still others, ΔH_f^0 and S_f^0 data are either available or can be calculated as f(T) so that ΔG_f^t can be calculated. The data for the new temperature can then be used to calculate the necessary Eh-pH boundaries using the Nernst equation as discussed earlier.

For many reactions, changes in pressure do not introduce significant errors in the Eh-pH boundaries calculated for 1 bar conditions. This is due to the fact that pressure increases on condensed phases (solids, aqueous species) are slight and usually in the same direction, so that the ΔG_R is close to ΔG_R^0. Basically, this approach must be used, while there are abundant data for temperatures other than 25 °C, there are few data for pressures other than 1 bar. Brookins (1979b) has discussed this problem in more detail.

Stability Limits for Water

The stability field for water at 25 °C and 1 bar pressure is defined by the reaction:

$$2H_2O(liq) = 2H_2(g) + O_2(g) \ .$$

This reaction is the sum of the two half-cells:

a) upper limit:

$$2H_2O = 4H^+ + O_2(g) + 4e^- \;.$$

b) lower limit:

$$H_2(g) = 2H^+ + 2e^- \;.$$

While water does not enter into half-cell b, it is implied by the presence of H^+.

The upper limit (a) is calculated based on the ΔG_f^0 data for the species involved and the Nernst equation. Since the ΔG_f^0 of any element in its standard state is zero, and since the ΔG_f of both H^+ and e^- are zero by convention, then only the $\Delta G_{f_{H_2O}}^0$ affects reaction a.

Thus, $\Delta G_R^0 = +113.4$ kcal for reaction a, and $E^0 = 1.23$ V. The Eh-pH equation is:

$$Eh = E^0 + \frac{0.059}{n} \log (P_{O_2}) + \frac{0.059}{n} \log (H^+)^4$$

and

$$Eh = 1.23 \text{ V} - 0.059 \text{ pH} \;.$$

Similarly, for reaction b, since $\Delta G_R^0 = 0$, then the Eh-pH expression is:

$$Eh = 0.00 - 0.059 \text{ pH} \;.$$

It is apparent from these equations for the stability limits of water that over 83 orders of magnitude for P_{O_2} and 41 orders of magnitude for P_{H_2} are covered. Thus, for the atmospheric conditions, where $P_{O_2} \sim 0.2$ bar, then the Eh-pH expression is:

$$Eh = 1.23 + \frac{0.059}{2} \log (0.2) - 0.059 \text{ pH}$$

$$Eh = 1.22 - 0.059 \text{ pH} \;.$$

Thus, the upper stability limit of water for $P_{O_2} = 1$ bar is very close to that boundary for $P_{O_2} \sim 0.2$ bar.

Natural Waters

Baas Becking et al. (1960) plotted many data from natural waters, and a modified version of his Eh-pH diagram is shown in Fig. 1. As pointed out by Krauskopf (1979), Baas Becking et al. (1960) included measurements from very unusual waters, such as extremely acidic oxidizing and reducing conditions and extremely basic oxidizing and reducing conditions. Krauskopf (1979) argues that most natural waters yield data that plot between the pH limits of roughly 4 to 9, and that this part of the diagram will be the most important for geochemists. In recent years, however, with the great amount of interest on environmental geochemistry, even the very acidic and very basic conditions are of interest. Waters from many mine and coal tailings are extremely acidic, as are a variety of waters accompanying chemical wastes. Similarly, alkaline leaching produces high pH waters with wastes, and high pH waters are known from certain tuffaceous rocks such as those in Nevada being considered for high level radioactive waste disposal. Thus, in this book all diagrams will include, where possible, data and boundaries for species stable under very acidic to very basic conditions. Further, in the text accompanying the Eh-pH diagram for each element, I will attempt to address such points in more detail.

Other information of interest is plotted in Fig. 1. A very broad and loose indication of oxidizing, transitional, and reducing environments is indicated, and specific (but still somewhat variable) waters are identified as well. These, identified by numbers, are: (1) mine waters, which are usually acidic and oxidizing; (2) rain, which in certain parts of the world is becoming much more acidic due to the buildup of CO_2 in the atmosphere, and in turn due to fossil fuel burning, is acidic and getting more so. (3) Stream waters are oxidizing and near-neutral in pH, although, as in the case for rain, increased acidity is noted due to acid rain and acid mine drainage, especially in the northeastern USA. (4) Normal ocean water, which in the near-surface to the surface is oxidizing and in approximate equilibrium with both $CO_2(g)$ and $CaCO_3(c)$ as well as dissolved CO_2 (i.e., $HCO_3^- + CO_3^{2-} + H_2CO_3$). (5) The residual aerated waters in saline environments are typically oxidizing and basic; in some cases the pH may be above 10. (6) Groundwaters in many instances are near saturation with respect to $CaCO_3$, and this is reflected in the approximate pH near 8.4. They are much less oxidizing than surface waters, and many reduction reactions occur in this zone as well. (7) Bog waters tend to be mildly reducing, and acidic. These cross the sulfide-sulfate boundary as shown in Fig. 1. (8) Water-logged soils are reducing and acidic, and authigenic minerals found in such environments clearly reflect these Eh-pH conditions. (9) Euxenic marine waters are reducing and near-neutral in pH. (10) Saline waters rich in organic material are reducing and alkaline. These numbered areas are redrawn from Garrels and Christ (1965), and the reader is referred to this source as well as to Krauskopf (1979), Drever (1982), and Brookins (1984) for additional general information on Eh-pH diagrams and their utilization.

Also shown in Fig. 1 are two very important boundaries in Eh-pH space. The first of these is the boundary between reduced and oxidized sulfur species. This sulfide-sulfate boundary separates overall chemically reducing conditions from mildly oxidizing conditions. The field of native sulfur is not shown here, and the reader is referred to the sulfur species Eh-pH diagram (Fig. 3) here as well as to Garrels and Christ (1965) for more detail. This boundary separates metal sulfides from their more oxidized sulfate-material (and carbonate-mineral) alteration products, for example. In the oxidation of $S(-II)$ in sulfide

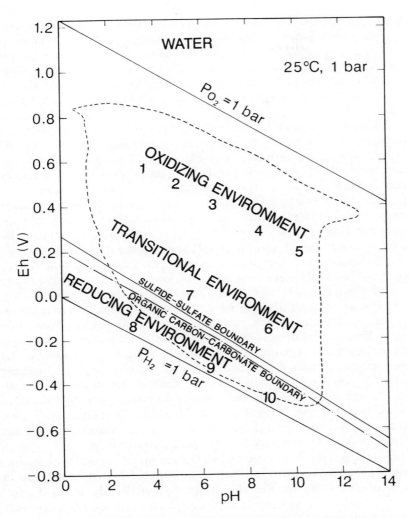

Fig. 1. Generic Eh-pH diagram for water showing several features.
The upper and lower stability limits of water are given where P_{O_2} and $P_{H_2} = 1$ bar, respectively. The *dotted line* represents the range of natural Eh-pH water measurements reported by Baas Becking et al. (1960). The *solid line* marked sulfide-sulfate boundary separates oxidizing, sulfate-bearing waters above this boundary to reducing, sulfide (as H_2S or HS^-)-bearing waters below. The *dash-dot* boundary separates organic carbon (below boundary) from carbonate species (above boundary). Oxidizing, transitional, and reducing environments are indicated. Specific environments are (*numbers on figure*): *1* mine waters; *2* rain; *3* streams; *4* normal ocean water; *5* aerated saline water residues; *6* ground waters; *7* bog waters; *8* water-logged soils; *9* euxenic marine waters; *10* organic-rich, saline waters.
See text for discussion

to S(VI) in sulfate, there is a transfer of eight electrons, hence this is an extremely energetic reaction, so much that at times certain metals, in attempting to counterbalance this oxidation, are reduced to lower valences. This is the case for copper, silver and the platinum group elements, for example. As sulfides are destroyed, even metals released without a change in valence, are commonly fixed in new oxysalts, and their recognition can be used for geochemical exploration.

The boundary for organic carbon oxidizing to carbonate is also shown, and this boundary, too, is extremely important in geochemistry. The Eh-pH diagram for C species is shown later (Fig. 13). Basically, $C(O)$ and/or $C(-IV)$ (i.e., in methane or similar material) is oxidized to $C(IV)$ in carbonic acid or carbonate minerals. Any metals released by destruction of coaly material, for example, may form carbonates, or, since such material is commonly pyritiferous, sulfate minerals penecontemporaneous with carbonates as well.

From the 98 figures shown in this book, the reader will note that many inorganic solid species occupy specific Eh-pH space. Since many minerals have fairly simple chemistry (i.e., they are readily identified by one major metal ion), recognition of the Eh-pH space occupied by various minerals allows the reader to determine which minerals are probably cogenetic, which may be metastable, and so on. Nature conveniently displays very repetitive mineralogy in many instances, and by identifying the minerals present and working out their paragenesis, Eh-pH conditions for the environment of formation can be determined. This has been proven to be successful in explaining retention versus migration of radioactive elements in natural occurrences, for example (Brookins 1984; on the Oklo natural reactor, Gabon), for uranium mill tailings, for hazardous wastes, for new phases forming from contaminated streams, and in many other places. Once this Eh-pH has been determined, then use of the same diagrams for aqueous species of the elements of interest can be inspected to attempt to determine transport characteristics.

Yet Eh-pH diagrams, as pointed out earlier, are limited in that they assume equilibrium processes and this has been demonstrated not to be the case in most of nature (Lindberg and Runnells 1984). Further, they do not yield information on kinetics of reactions. Thirdly, complexities of solid solution may make the Eh-pH area of an one metal-dominant species questionable. Yet despite these drawbacks, these diagrams have been demonstrated to be extremely useful to the geochemist time and time again, and it is my hope that the diagrams presented here will make this job of determining mineral paragenesis, and/or of studying different kinds of waste disposal, easier.

Presentation of the Elements and Elements Covered

The standard order of arrangement of the elements is advocated in this book. This follows the usage of the US National Bureau of Standards. In such treatment the order of the elements is from oxygen to hydrogen to the inert gases to the halides and so on to the alkaline earth and alkali elements. This standard order is shown in Wagman et al. (1982; Fig. 1).

For this book the following elements are not presented in terms of Eh-pH diagrams: oxygen, hydrogen, helium, neon, argon, krypton, xenon, radon, fluorine, chlorine, bromine, astatine, promethium, curium and above transuranics, protoactinium, actinium, lithium, sodium, potassium, rubidium, cesium, francium. Since all of the Eh-pH diagrams are presented in terms of the stability limits of water at 25 °C and 1 bar total pressure, both P_{O_2} and P_{H_2} are defined by these limits. The inert gases are not chemically reactive under normal conditions and thus are not treated here. Of the halogens, only iodine shows more than one stable oxidation state in the water stability field (Fig. 2), while the others are present over the entire stability field of water as M^{1-} species. Promethium contains no long-lived stable isotopes in nature, hence too few thermodynamic data are available for this element for Eh-pH calculations. This is also the case for curium and the higher transuranics, as well as for protoactinium and actinium. The alkali elements occur in nature as M^{1+} ions over the entire water stability field and thus are not presented.

All the remaining elements are presented in Eh-pH diagrams in this book. This includes elements with only radioactive isotopes such as technetium, polonium, thorium, uranium, neptunium, plutonium, and americium for which, in part due to the extreme interest in these elements in conjunction with radioactive waste diposal and related fields, adequate data bases are available.

Thermodynamic data for water and its constituents are given in Table 1 since these data are used in all subsequent tables and calculations.

Table 1. Thermodynamic data for water

Species (state)	ΔG_f^0 (kcal/gfw)	Reference
H_2O (liq)	−56.69	Wagman et al. (1982)
OH^- (aq)	−37.59	Wagman et al. (1982)
H^- (aq)	0.00	Wagman et al. (1982)
H_2 (g)	0.00	Wagman et al. (1982)
O_2 (g)	0.00	Wagman et al. (1982)
H_2O (g)	−54.63	Wagman et al. (1982)

Abbreviations for Tables 1 − 61

liq = liquid	am	= amorphous
aq = aqueous	a-qz	= alpha-quartz
g = gaseous	c, brown	= crystalline, brown
c = crystalline	c, red	= crystalline, red

gibbsite, anatase, calcite, aragonite, anhydrite, gypsum, mineral names

IODINE AND OTHER HALIDES

The Eh-pH diagram for iodine species is shown in Fig. 2. The thermodynamic data for iodine and other halides are given in Table 2. The halides form important anions in the terrestrial environment, and naturally occurring chlorides, fluorides, iodides, and bromides occur. Most of these are soluble in water, such as halite (NaCl) and sylvite (KCl).

Of the halides only iodine possesses more than one oxidation state in the limits of the stability field of water. The boundary between I and IO_3^- has a slope of -0.059 pH and intersects the Eh axis at 1.09 V. The same boundaries for $Cl^- : ClO_3^-$ and $Br^- : BrO_3^-$ fall above the upper stability limit of water, intersecting the Eh axis at 1.44 V for both. Data for FO_3^- are lacking. Thus, the stable halides in water are Cl^-, Br^-, F^-, I^-, and IO_3^-.

The thermodynamic data for the halides are all taken from Wagman et al. (1982) and are internally consistent.

Higher oxidation states of the halides (M VII) all plot well above the upper stability field of water.

Table 2. Thermodynamic data for iodine and other halides

Species (state)	ΔG_f^0 (kcal/gfw)	Reference
F^- (aq)	-66.63	Wagman et al. (1982)
Cl^- (aq)	-31.36	Wagman et al. (1982)
ClO_3^- (aq)	-1.90	Wagman et al. (1982)
ClO_4^- (aq)	-2.04	Wagman et al. (1982)
Br^- (aq)	-24.85	Wagman et al. (1982)
BrO_3^- (aq)	$+4.46$	Wagman et al. (1982)
BrO_4^- (aq)	$+28.23$	Wagman et al. (1982)
I^- (aq)	-12.33	Wagman et al. (1982)
IO_3^- (aq)	-30.59	Wagman et al. (1982)
IO_4^- (aq)	-13.98	Wagman et al. (1982)

Abbreviations see Table 1

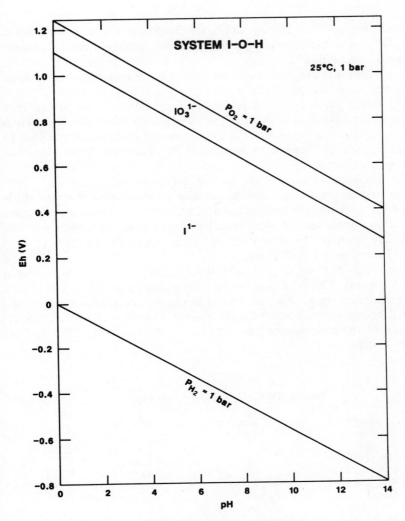

Fig. 2. Eh-pH diagram for part of the system I-O-H. See text for discussion

SULPHUR

The Eh-pH diagram for sulfur species in the water stability field is shown in Fig. 3. The thermodynamic data for important naturally occurring sulfur species are given in Table 3.

The sulfur Eh-pH diagram presented here is calculated on the basis of an assumed activity of dissolved sulfur = 10^{-3}, since this is a reasonable value for many terrestrial waters. Garrels and Christ (1965) and others have presented sulfur Eh-pH diagrams for both much greater and lower activities of dissolved sulfur. For an increase in the activity of dissolved S, the field of native sulfur increases, and for a decrease in the S activity, it decreases and disappears at about 10^{-6}.

Sulfur is an important element in the terrestrial systeme. Here, I have chosen to include only those species of major importance, and not those species of short, metastable existence. Thus, sulfites and thiosulfites and other anionic sulfur species are not given here. The diagrams for various metals with sulfur in the form of metal sulfides are given under the heading of each metallic element later in this book.

The boundary between $S(-II)$ and $S(VI)$ is one of the most important in terrestrial systems. The oxidation of $S(-II)$ species to $S(VI)$ species is very energetic, and important for a wide variety of changes in mineral assemblages. These will be commented on under the various metallic element headings.

The thermodynamic data are all taken from Wagman et al. (1982) and are internally consistent.

Table 3. Thermodynamic data for sulfur

Species (state)	ΔG_f^0 (kcal/gfw)	Reference
S^{2-} (aq)	+20.51	Wagman et al. (1982)
HS^- (aq)	+2.89	Wagman et al. (1982)
H_2S (aq)	−6.65	Wagman et al. (1982)
SO_3^{2-} (aq)	−116.28	Wagman et al. (1982)
HSO_3^- (aq)	−126.13	Wagman et al. (1982)
SO_4^{2-} (aq)	−177.95	Wagman et al. (1982)
HSO_4^- (aq)	−180.67	Wagman et al. (1982)

Abbreviations see Table 1

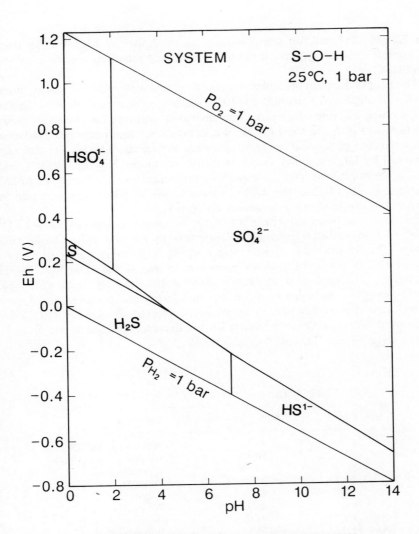

Fig. 3. Eh-pH diagram for part of the system S-O-H.
The activity of dissolved S = 10^{-3} (roughly 32 ppm) for convenience. The importance of the boundary between sulfate (HSO_4^-, SO_4^{2-}) and sulfide (H_2S, HS^-, S^{2-}), exclusive of native S, is discussed in the text

SELENIUM

The Eh-pH diagram for selenium species is shown in Fig. 4. The thermodynamic data for important naturally occurring selenium species are given in Table 4.

The Eh-pH diagram presented here is calculated on the basis of an assumed activity of dissolved selenium of 10^{-6}. Selenium is a rare element in the earth's crust, but one of considerable environmental interest due to its ready incorporation into the food chain. While excess selenium can cause adverse health effects, so, too, can extreme selenium deficiency. The Eh-pH diagram is marked by large stability fields of native selenium, $HSeO_3^-$, SeO_3^{2-}, and SeO_4^{2-}. The selenite species are especially important as they have been linked to adverse health effects. The field of native Se will increase with increased Se activity, and diminish with decreased Se activity.

Selenium can often substitute for sulfur in nature, commonly as Se $(-II)$ for S$(-II)$ in sulfides, forming seleniferous pyrite and other Se-bearing sulfides. The Se analog for pyrite, ferroselite ($FeSe_2$) occurs rarely in nature. Howard (1977) has published an extensive review of the selenium Eh-pH systematics, and presented diagrams showing the stability field of $FeSe_2$. There is not good agreement on the ΔG_f^0 for ferroselite, however, and I have not included the species here. The reader is referred to Howard (1977) for details on this matter. Other selenium Eh-pH diagrams of interest are given in Brookins (1979) and Thomson et al. (1986).

Table 4. Thermodynamic data for selenium

Species (state)	ΔG_f^0 (kcal/gfw)	Reference
Se^{2-} (aq)	+30.90	Wagman et al. (1982)
HSe^- (aq)	+10.50	Wagman et al. (1982)
H_2Se (aq)	+3.80	Wagman et al. (1982)
H_2SeO_3 (aq)	−101.85	Wagman et al. (1982)
$HSeO_3^-$ (aq)	−98.34	Wagman et al. (1982)
SeO_3^{2-} (aq)	−88.38	Wagman et al. (1982)
$HSeO_4^-$ (aq)	−108.08	Wagman et al. (1982)
SeO_4^{2-} (aq)	−105.47	Wagman et al. (1982)

Abbreviations see Table 1

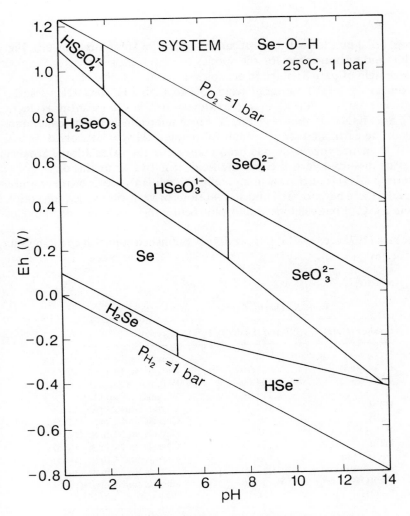

Fig. 4. Eh-pH diagram for part of the system Se-O-H. The activity of dissolved Se = 10^{-6}. See text for discussion

TELLURIUM

The Eh-pH diagram for naturally occurring tellurium species is shown in Fig. 5. The thermodynamic data for the species are given in Table 5.

A large field of native tellurium occupies the reducing Eh-pH space of Fig. 5. It oxidizes to Te(IV) forming, from low to high pH, $TeO \cdot OH^+$, TeO_2, $HTeO_3^-$, and TeO_3^{2-}. Te(IV) in turn oxidizes to Te(VI) to form H_2TeO_4, $HTeO_4^-$, and TeO_4^{2-} as shown in Fig. 5. Since tellurium is an extremely rare element in the earth's crust, it is doubtful if the activity of dissolved Te will exceed 10^{-6}, hence this value has been chosen for the calculations. Smaller activities of dissolved Te will cause the field of native Te to shrink.

Tellurium (−II) is also known, and telluride minerals are known in nature. Presumably, these have formed under conditions of very low oxygen fugacities, and their 25 °C, 1 bar analogs would plot below the lower stability limit of water.

Brookins (1978 a, c; 1987 a) has previously published some Eh-pH diagrams for Te species.

Table 5. Thermodynamic data for tellurium

Species (state)	ΔG_f^0 (kcal/gfw)	Reference
H_2Te (aq)	+34.10	Garrels and Christ (1965)
HTe^- (aq)	+37.70	Garrels and Christ (1965)
TeO_2 (c)	−64.60	Wagman et al. (1982)
H_2TeO_3 (aq)	−113.84	Wagman et al. (1982)
$HTeO_3^-$ (aq)	−104.34	Garrels and Christ (1965)
TeO_3^{2-} (aq)	−93.79	Garrels and Christ (1965)
H_2TeO_4 (aq)	−131.66	Garrels and Christ (1965)
$HTeO_4^-$ (aq)	−123.27	Garrels and Christ (1965)
TeO_4^{2-} (aq)	−109.10	Garrels and Christ (1965)
Te^{4+} (aq)	+52.38	Garrels and Christ (1965)
$TeO(OH)^+$ (aq)	−61.78	Garrels and Christ (1965)
$Te(OH)_3^+$ (aq)	−118.57	Wagman et al. (1982)

Abbreviations see Table 1

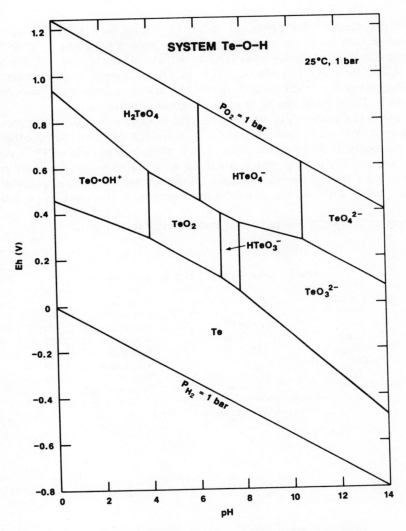

Fig. 5. Eh-pH diagram for part of the system Te-O-H. The activity of dissolved Te = 10^{-6}. See text for discussion

POLONIUM

The Eh-pH diagram for polonium species is shown in Fig. 6. The thermodynamic data for some important polonium species are given in Table 6.

Polonium isotopes are all radioactive, and ^{208}Po has the longest half-life of these at 2.9×10^3 years. In addition, ^{210}Po (half-life 138 days) is an intermediate decay product of ^{222}Rn (in turn from ^{238}U) and an is important potential carcinogen. Thus, polonium is of environmental importance.

The Eh-pH diagram (Fig. 6) shows a large field of native polonium, and a somewhat isolated field of PoS. PoO$_2$ is important under oxidizing conditions, especially at intermediate to basic pH, while a large field of Po^{2+} occupies most of the acidic, oxidizing Eh-pH space. The assumed activity of dissolved Po = 10^{-8} is reasonable based on the few polonium data available. A small field of Po^{4+} occurs under the most extreme acidic, oxidizing conditions.

Brookins (1978 b, c) has published earlier versions of Fig. 6.

Table 6. Thermodynamic data for polonium

Species (state)	ΔG_f^0 (kcal/gfw)	Reference
Po^{2+} (aq)	+16.97	Wagman et al. (1982)
Po^{4+} (aq)	+70.00	Wagman et al. (1982)
Po(OH)$_4$ (c)	−130.00	Wagman et al. (1982)
Po(OH)$_2^{2+}$ (aq)	−113.05	Wagman et al. (1982)
PoS (c)	−0.96	Wagman et al. (1982)
PoO$_2$ (c)	−46.60	Latimer (1952)

Abbreviations see Table 1

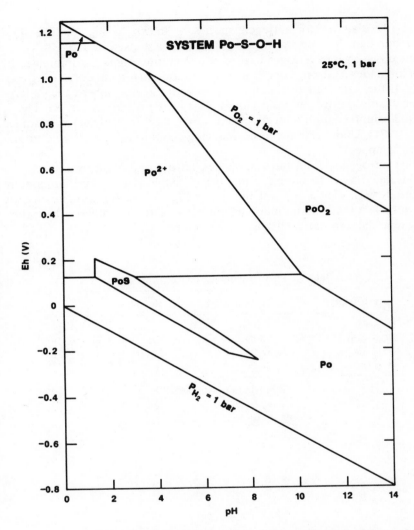

Fig. 6. Eh-pH diagram for part of the system Po-S-O-H. The assumed activities are: Po = 10^{-8}, S = 10^{-3}

NITROGEN

The Eh-pH diagram for nitrogen species is shown in Fig. 7. The thermodynamic data for important nitrogen species are given in Table 7.

Aqueous species of nitrogen in surface and groundwaters are dominated by nitrate ion under oxidizing conditions and ammonium ion under reducing conditions. If the waters are in communication with the atmosphere, then a large field of dissolved nitrogen gas occupies most of the Eh-pH space. For Fig. 7, the equilibrium condition of $P_{N_2} = 0.8$ atm ($= 10^{-3.3}$ activity) is assumed (Berner 1971). Under extreme basic, reducing conditions, a small field of ammonia gas appears.

Other nitrogen species, nitrites, nitriles, nitrous and nitric oxides, etc., are not considered here. In nature, a number of stable nitrates and ammonium compounds exist. These are not plotted in Fig. 7 because most nitrites are water soluble as are ammonium halide salts, and, for others, their thermodynamic data are not well known.

Table 7. Thermodynamic data for nitrogen

Species (state)	ΔG_f^0 (kcal/gfw)	Reference
N_2 (g)	0.00	Wagman et al. (1982)
NO_3^- (aq)	−25.99	Wagman et al. (1982)
NH_4^+ (aq)	−18.96	Wagman et al. (1982)
NH_3 (g)	−3.98	Berner (1971)

Abbreviations see Table 1

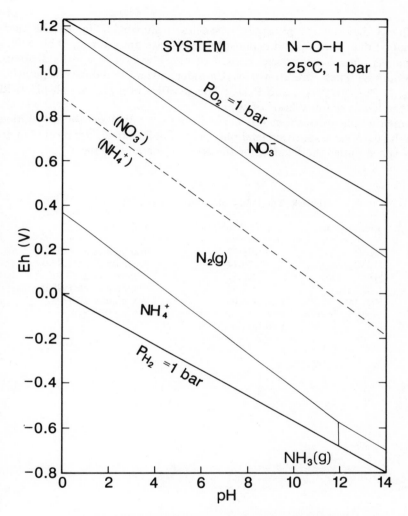

Fig. 7. Eh-pH diagram for part of the system N-O-H. The assumed activity of dissolved nitrogen = $10^{-3.3}$ (P_{N_2} = 0.8 bar). See text for discussion

PHOSPHORUS

The Eh-pH diagram for phosphorus species is shown in Fig. 8. The thermodynamic data for important phosphorus species are given in Table 8.

Aqueous phosphorus species consist of phosphoric acid and its dissociation products. These are shown as Eh-independent boundaries separating H_3PO_4, $H_2PO_4^-$, HPO_4^{2-}, and PO_4^{3-}. The gas PH_3 occupies an Eh-pH field below the lower stability limit of water.

Numerous phosphate minerals occur in nature, but these are not plotted here. Where relevant, specific metal phosphate stability fields are plotted in an Eh-pH space under the heading of the specific metallic element.

Table 8. Thermodynamic data for phosphorus

Species (state)	ΔG_f^0 (kcal/gfw)	Reference
H_3PO_4 (aq)	−273.07	Wagman et al. (1982)
$H_2PO_4^-$ (aq)	−270.14	Wagman et al. (1982)
HPO_4^{2-} (aq)	−260.31	Wagman et al. (1982)
PO_4^{3-} (aq)	−243.48	Wagman et al. (1982)
PH_3 (aq)	+6.06	Wagman et al. (1982)

Abbreviations see Table 1

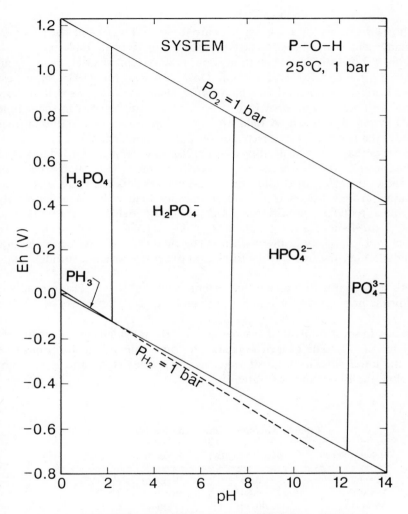

Fig. 8. Eh-pH diagram for part of the system P-O-H. The assumed activity of dissolved P = 10^{-4}. See text for discussion

ARSENIC

The Eh-pH diagrams for arsenic species are shown in Figs. 9 and 10. The thermodynamic data for important arsenic species are given in Table 9.

Figure 9 shows the Eh-pH relationships in the system As-O-H assuming an activity of 10^{-6} for dissolved arsenic. Dove and Rimstidt (1985) have also shown a similar diagram and, in a later figure, also included Fe as well to show the stability field of scorodite. In Fig. 9, however, only species in the simple As-O-H system are shown. A field of As_2O_3 is indicated by the dotted lines just above the field of native arsenic under mildly reducing, acidic conditions. Figure 9 is perhaps a bit misleading as it implies a moderate to fairly large field of native arsenic, and this occurrence is rare in nature. Figure 10, on the other hand, shows the effect of sulphur on the arsenic systematics. Here, it is noted that native arsenic occurs in the water stability field only under extremely basic, most reducing conditions. In Fig. 10, the fields of arsenous acid (H_3AsO_3) and its ionization products ($H_2AsO_3^-$, $HAsO_3^{2-}$, and AsO_3^{3-}) have been omitted for clarity in presentation. The field of As_2O_3 (either arsenolite or claudetite) occurs under acidic to neutral pH just above the sulfide-sulfate fence.

Both Figs. 9 and 10 indicate that arsenic acid and its ionization products are of prime importance for As transport under a very wide range of Eh and pH.

Arsenic in some minerals is present as As($-$II), in which case it substitutes readily for S($-$II). The oxygen fugacities for such minerals to form must be below the lower stability limit of water as the species H_2As and HAs^- plot well below the lower limit of water.

Table 9. Thermodynamic data for arsenic

Species (state)	ΔG_f^0 (kcal/gfw)	Reference
As (c)	0.00	Wagman et al. (1982)
AsS (c)	-16.80	Robie et al. (1978)
As_2S_3 (c)	-40.30	Wagman et al. (1982)
As_2O_3 (c)	-137.66	Robie et al. (1978)
H_3AsO_4 (aq)	-183.08	Wagman et al. (1982)
$H_2AsO_4^-$ (aq)	-180.01	Wagman et al. (1982)
$HAsO_4^{2-}$ (aq)	-170.69	Wagman et al. (1982)
AsO_4^{3-} (aq)	-154.97	Wagman et al. (1982)
H_3AsO_3 (aq)	-152.92	Wagman et al. (1982)
$H_2AsO_3^-$ (aq)	-140.33	Wagman et al. (1982)
$HAsO_3^{2-}$ (aq)	-125.31	Dove and Rimstidt (1985)
AsO_3^{3-} (aq)	-107.00	Dove and Rimstidt (1985)

Abbreviations see Table 1

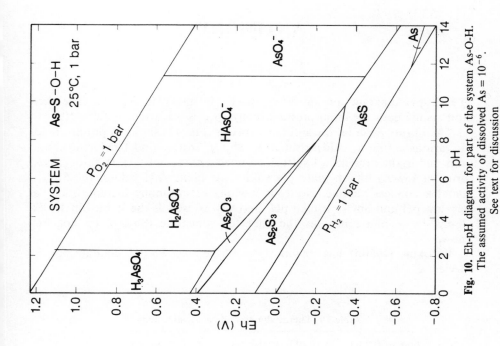

Fig. 10. Eh-pH diagram for part of the system As-S-O-H. The assumed activity of dissolved As = 10^{-6}. See text for discussion

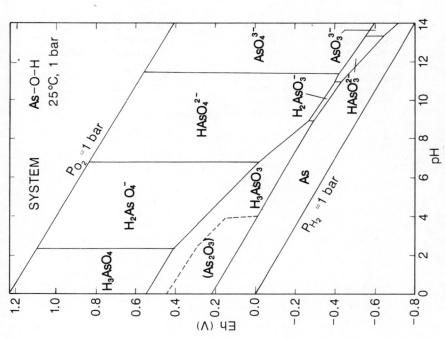

Fig. 9. Eh-pH diagram for part of the system As-S-O-H. The assumed activities of dissolved species are: As = 10^{-6}, S = 10^{-3}. See text for discussion

ANTIMONY

The Eh-pH diagram for antimony species is shown in Fig. 11. The thermodynamic data for important antimony species are given in Table 10.

The Eh-pH diagram for antimony species (Fig. 11) shows important fields for various antimony oxides, $Sb(OH)_3$, Sb_2O_4, Sb_2O_5. Under reducing conditions and in the presence of sulfur (assumed activity = 10^{-3}), stibnite (Sb_2S_3) forms at low to intermediate pH values. At higher pH values, the $Sb_2S_4^{2-}$ complex replaces stibnite. Aqueous oxyions of antimony include SbO^+ at very low pH and SbO_2^- at high pH. $Sb(OH)_3$ is used as the stable Sb(III) oxide-hydroxide phase here as the data for stibconite (Sb_2O_3) are not well known.

Brookins (1986a) has previously published an Eh-pH diagram for Sb species.

Table 10. Thermodynamic data for antimony

Species (state)	ΔG_f^0 (kcal/gfw)	Reference
SbO^+ (aq)	−42.33	Wagman et al. (1982)
SbO_2^- (aq)	−81.31	Wagman et al. (1982)
Sb_2S_3 (c)	−41.49	Wagman et al. (1982)
$Sb_2S_4^{2-}$ (aq)	−23.78	Wagman et al. (1982)
$HSbO_2$ (aq)	−97.39	Wagman et al. (1982)
$Sb(OH)_3$ (c)	−163.77	Wagman et al. (1982)
Sb_2O_4 (c)	−190.18	Wagman et al. (1982)
Sb_2O_5 (c)	−198.18	Wagman et al. (1982)

Abbreviations see Table 1

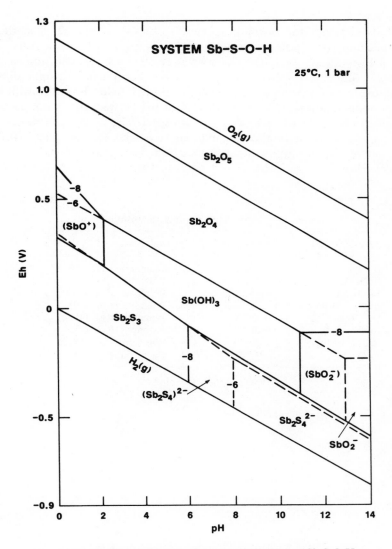

Fig. 11. Eh-pH diagram for part of the system Sb-S-O-H. The assumed activities of dissolved Sb = $10^{-6,-8}$; S = 10^{-3}. See text for discussion

BISMUTH

The Eh-pH diagram for bismuth species is shown in Fig. 12. The thermodynamic data for important bismuth species are given in Table 11. Aqueous species of bismuth are complex. The species $Bi_6O_6^{6+}$ is stable with respect to Bi^{3+}, BiO^+, and $BiOH^{2+}$ based on the avialable thermodynamic data (Table 11). The oxide Bi_2O_3 occupies much of the Eh-pH field above the sulfide-sulfate boundary, and is replaced by $Bi_6O_6^{6+}$ at pH between 5.5 and 6 for activities of dissolved Bi of 10^{-6} and 10^{-8}, respectively. Under sulfide-stable conditions, Bi_2S_3, bismuthinite, occupies most of the Eh-pH space with a field of native bismuth appearing at high pH, reducing conditions.

Earlier Eh-pH diagrams have been published by Brookins (1978 b, c).

Table 11. Thermodynamic data for bismuth

Species (state)	ΔG_f^0 (kcal/gfw)	Reference
Bi^0 (c)	0.00	Wagman et al. (1982)
Bi_2S_3 (c)	-33.60	Wagman et al. (1982)
Bi_2O_3 (c)	-118.00	Wagman et al. (1982)
$Bi_6O_6^{6+}$ (aq)	-221.80	Wagman et al. (1982)
Bi^{3+} (aq)	$+19.79$	Wagman et al. (1982)
BiO^{1+} (aq)	-34.99	Latimer (1952)
$BiOH^{2+}$ (aq)	-34.99	Wagman et al. (1982)
Bi_4O_7 (c)	-232.75	Garrels and Christ (1965)

Abbreviations see Table 1

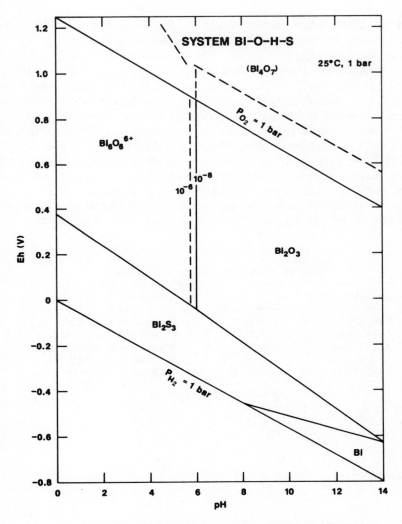

Fig. 12. Eh-pH diagram for part of the system Bi-O-H-S. The assumed activities of dissolved species are: $Bi = 10^{-6-8}$, $S = 10^{-3}$. See text for discussion

CARBON

The Eh-pH diagram for carbon species is shown in Fig. 13. The thermodynamic data for important carbon species are given in Table 12.

Of the extremely large number of carbon species known in nature, I have chosen to consider only those species shown in Fig. 13. Under oxidizing conditions at all pH values, carbonic acid and its ionization products dominate. Carbonic acid is the important species below pH = 6.4, bicarbonate from 6.4 to 10.3, and carbonate ion above pH = 10.3. As reducing conditions are encountered, and very close to the sulfide-sulfate boundary, C(IV) is reduced to C(0) or C(−IV) as indicated. While the Eh-pH diagram shown here is very much simplified, it should be noted that most organic carbon species with C(0) fall in the general range of native carbon shown in Fig. 13. Aqueous methane is shown as the important species of C(−IV) for convenience.

There are numerous carbonate minerals known in nature, and these are presented in the Eh-pH diagrams for various metals considered herein.

Table 12. Thermodynamic data for carbon

Species (state)	ΔG_f^0 (kcal/gfw)	Reference
CH_4 (aq)	−8.28	Wagman et al. (1982)
CH_2O (aq)	−31.00	Garrels and Christ (1965)
H_2CO_3 (aq)	−149.00	Wagman et al. (1982)
HCO_3^- (aq)	−140.24	Wagman et al. (1982)
CO_3^{2-} (aq)	−126.15	Wagman et al. (1982)

Abbreviations see Table 1

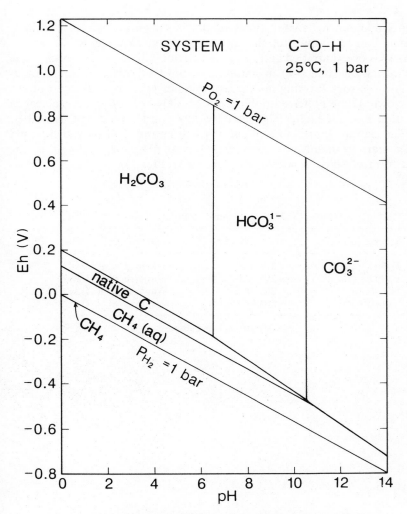

Fig. 13. Eh-pH diagram for part of the system C-O-H.
The assumed activity of dissolved C = 10^{-3}. See text for discussion

SILICON

The Eh-pH diagram for some silicon species is shown in Fig. 14. The thermodynamic data for important silicon species are given in Table 13.

Figure 14 is oversimplified and somewhat trivial. Silicon possesses only one common oxidation state (IV) in nature, and most of this element is incorporated into the rock-forming and accessory silicates. In Fig. 14 amorphous silica is shown as the preferred polymorph of SiO_2 as it often controls silica precipitation, but, for higher SiO_2 activity, low quartz could just as well be chosen. The extreme insolubility of silica is well known, and only at high pH does silica start to dissolve to form $H_3SiO_4^-$. At pH values below this, the dissolved form of silica is H_4SiO_4 (not shown in Fig. 14).

Table 13. Thermodynamic data for silicon

Species (state)	ΔG_f^0 (kcal/gfw)	Reference
H_4SiO_4 (aq)	−314.53	Wagman et al. (1982)
$H_3SiO_4^-$ (aq)	−295.53	Wagman et al. (1982)
SiO_2 (am)	−203.32	Wagman et al. (1982)
SiO_2 (a−qz)	−204.70	Wagman et al. (1982)

Abbreviations see Table 1

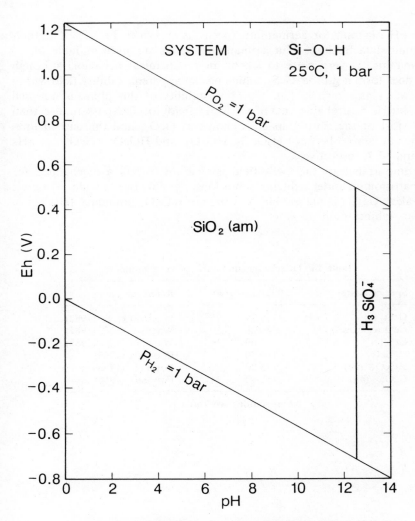

Fig. 14. Eh-pH diagram for part of the system Si-O-H. The assumed activity of dissolved Si = 10^{-3}. See text for discussion

GERMANIUM

The Eh-pH diagram for germanium species is shown in Fig. 15. The thermodynamic data for important germanium species are given in Table 14.

Germanium is very close to silicon in geochemical behavior, although Ge(IV) does form a sulfide, GeS_2, which has an apparent stability field within that of water as shown in Fig. 15. The solubility of this phase is not well known, but it is probable that the Eh-pH field for GeS_2 is smaller than shown. Most of the Eh-pH space is shown as GeO_2, and the dashed lines show the boundaries between $H_2GeO_3 : HGeO_3^-$ and $HGeO_3^- : GeO_3^{2-}$ at pHs of 8.5 and 12.7, respectively.

The importance of the GeS_2 field is that this offers a mechanism for Se/Ge partitioning under reducing conditions, possibly over a wide pH range. The divalent Ge(II) is metastable, and the $GeO : GeO_2$ boundary falls below the lower stability limit of water.

Table 14. Thermodynamic data for germanium

Species (state)	ΔG_f^0 (kcal/gfw)	Reference
GeO_2 (c)	−118.79	Wagman et al. (1982)
GeO (c, brown)	−56.69	Wagman et al. (1982)
GeS (c)	−17.09	Wagman et al. (1982)
H_2GeO_3 (aq)	−186.80	Pourbaix (1966)
$HGeO_3^-$ (aq)	−175.20	Pourbaix (1966)
GeO_3^{2-} (aq)	−157.90	Pourbaix (1966)

Abbreviations see Table 1

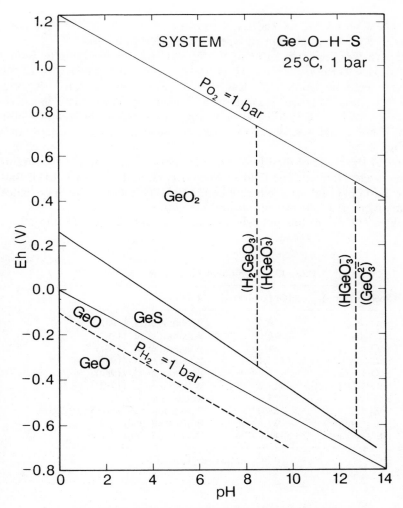

Fig. 15. Eh-pH diagram for part of the system Ge-O-H-S.
The assumed activities of dissolved species are: Ge = 10^{-6}, S = 10^{-3}.
See text for discussion

TIN

The Eh-pH diagram for tin species is shown in Figs. 16 and 17. The thermodynamic data for important tin species are given in Table 15.

In Fig. 16 it is seen that most of the Eh-pH space is occupied by SnO_2, which dissolves to SnO_3^{2-} above pH 11.9 ($a_{Sn} = 10^{-8}$) or 12.95 ($a_{Sn} = 10^{-6}$). Under very extreme acidic pH, less than 0, SnO_2 will dissolve to form $SnO \cdot OH^+$ or, less likely, Sn^{4+}. Divalent Sn(II) species are rare. For an activity of 10^{-8}, a very small field of $HSnO_2^-$ appears at the lower stability limit of water at about pH = 8, but SnO is entirely metastable. So, too, are Sn^{2+} and $SnOH^+$.

In Fig. 17 the effect of dissolved sulfur (activity = 10^{-3}) on the system Sn-S-O-H are seen. Here, a large field of SnS_2 occurs, and a small field of SnS occurs at basic pH and very reducing conditions. This shows the importance of sulfur on tin geochemistry in natural systems.

Brookins (1978a, c) has published earlier versions of Figs. 16 and 17.

Table 15. Thermodynamic data for tin

Species (state)	ΔG_f^0 (kcal/gfw)	Reference
Sn^{2+} (aq)	−6.50	Wagman et al. (1982)
Sn^{4+} (aq)	+0.60	Wagman et al. (1982)
SnO (c)	−61.40	Wagman et al. (1982)
SnO_2 (c)	−124.19	Wagman et al. (1982)
$SnOH^+$ (aq)	−60.90	Wagman et al. (1982)
$SnO \cdot OH^+$ (aq)	−113.29	Wagman et al. (1982)
SnS (c)	−25.02	Robie et al. (1978)
SnS_2 (c)	−38.00	Barner and Scheuerman (1978)
SnO_3^{2-} (aq)	−137.42	Pourbaix (1966)
$HSnO_2^-$ (aq)	−48.00	Pourbaix (1966)

Abbreviations see Table 1

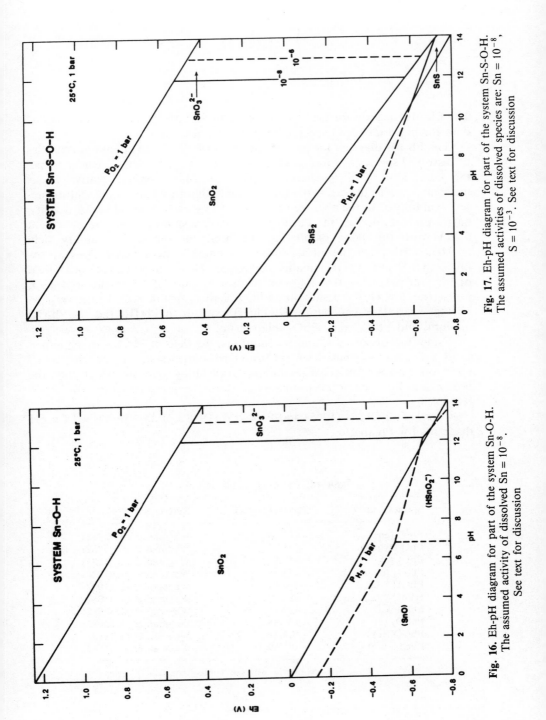

Fig. 17. Eh-pH diagram for part of the system Sn-S-O-H. The assumed activities of dissolved species are: Sn = 10^{-8}, S = 10^{-3}. See text for discussion

Fig. 16. Eh-pH diagram for part of the system Sn-O-H. The assumed activity of dissolved Sn = 10^{-8}. See text for discussion

LEAD

The Eh-pH diagram for lead species is shown in Fig. 18. The thermodynamic data for important lead species are given in Table 16.

The Eh-pH diagram for part of the system Pb-S-C-O-H assumes activities of dissolved species as follows: $Pb = 10^{-6}$, $S = 10^{-3}$, $C = 10^{-3}$. Under reducing conditions, PbS (galena) occupies the Eh-pH space. Native Pb is metastable, as it falls below the lower stability limit of water under sulfur-present conditions. In the absence of sulfur, or at greatly reduced sulfur activity, native Pb is stable [i.e., the Pb:PbO boundary would intersect the Eh axis at 0.26 V with the slope parallel to the limits of the water stability field (-0.059 pH)]. When the sulfide-sulfate boundary is encountered, S($-$II) is oxidized to S(VI) but Pb(II) remains unchanged. This results in a very small field of Pb^{2+} at pH below 0.4, followed (with increasing pH) by fields of $PbSO_4$ (anglesite), $PbCO_3$ (cerussite), and PbO (massicot or litharge). Under very oxidizing, near-neutral to basic pH, Pb(II) oxidizes to Pb(III) (i.e., in Pb_3O_4; minium) and Pb(IV) (i.e., PbO_2; plattnerite).

Under low dissolved C and S conditions, the field of Pb^{2+} is much larger. Lead is a very problematic element from environmental viewpoints. Tetraethyl lead from gasoline fumes causes buildup of Pb along roadways which enter the food chain; Pb in paint has been banned from use inside structures for health reasons.

Garrels and Christ (1965) and Brookins (1978 b, c) have published Eh-pH diagrams for Pb species.

Table 16. Thermodynamic data for lead

Species (state)	ΔG_f^0 (kcal/gfw)	Reference
Pb^{2+} (aq)	-5.84	Wagman et al. (1982)
PbO (c, red)	-45.16	Wagman et al. (1982)
PbO_2 (c)	-51.94	Wagman et al. (1982)
Pb_3O_4 (c)	-143.69	Wagman et al. (1982)
$PbOH^+$ (aq)	-54.09	Wagman et al. (1982)
$HPbO_2^-$ (aq)	-80.88	Wagman et al. (1982)
PbS (c)	-23.59	Wagman et al. (1982)
$PbSO_4$ (c)	-194.35	Wagman et al. (1982)
$PbCO_3$ (c)	-149.50	Wagman et al. (1982)
$PbCl_3^-$ (aq)	-101.69	Wagman et al. (1982)

Abbreviations see Table 1

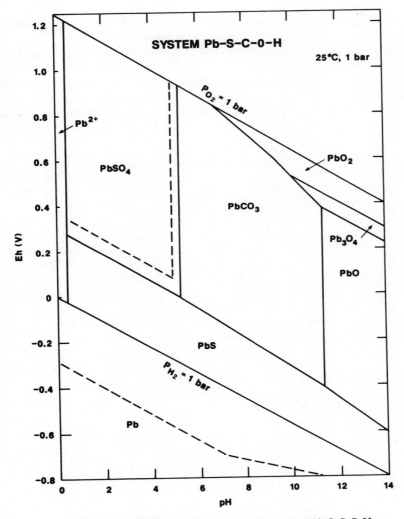

Fig. 18. Eh-pH diagram for part of the system Pb-S-C-O-H. The assumed activities of dissolved species are: Pb = $10^{-6,-8}$, S = 10^{-3}, C = 10^{-3}. See text for discussion

BORON

The Eh-pH diagram for boron species is shown in Fig. 19. The thermodynamic data for important boron species are given in Table 17.

The Eh-pH diagram for the system B-O-H (Fig. 19) shows a very large field of boric acid (H_3BO_3) and, above pH 10, its dissociation products. Data are given for a number of other B species in Table 17, but these are either outside the water limits (i.e., BH_4^-), water soluble (i.e., B_2O_3), or metastable with respect to boric acid [i.e., $B_4O_7^{2-}$, BO_2^-, $B(OH)_4^-$] or its ionization products.

Table 17. Thermodynamic data for boron

Species (state)	ΔG_f^0 (kcal/gfw)	Reference
BO_2^- (aq)	−162.26	Wagman et al. (1982)
B_2O_3 (c)	−285.29	Wagman et al. (1982)
$B_4O_7^{2-}$ (aq)	−622.56	Wagman et al. (1982)
BH_4^- (aq)	+27.33	Wagman et al. (1982)
$B(OH)_4^-$ (aq)	−275.61	Wagman et al. (1982)
H_3BO_3 (aq)	−231.54	Wagman et al. (1982)
$H_2BO_3^-$ (aq)	−217.60	Pourbaix (1966)
HBO_3^{2-} (aq)	−200.29	Pourbaix (1966)
BO_3^{3-} (aq)	−181.48	Pourbaix (1966)

Abbreviations see Table 1

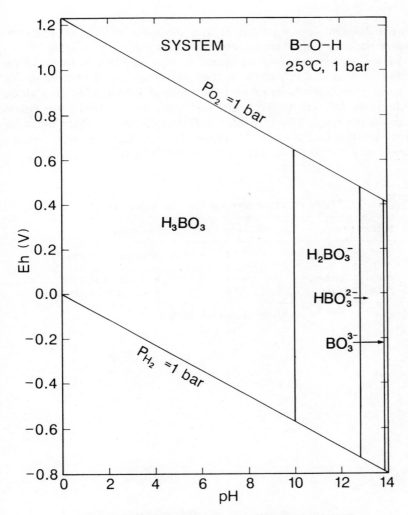

Fig. 19. Eh-pH diagram for part of the system B-O-H.
See text for discussion

ALUMINUM

The Eh-pH diagram for aluminum species is shown in Fig. 20. The thermodynamic data for important aluminum species are given in Table 18.

Aluminum possess only one important valence, Al(III) in nature, and the insolubility of Al over much of the natural pH range is well known. In Fig. 20 the stability fields as a function of pH are plotted. Under acidic conditions, Al is soluble as Al^{3+} or $AlOH^{2+}$, although only Al^{3+} is shown for convenience. In the pH range, 3.7–11.3 solid $Al(OH)_3$ (gibbsite) is stable, assuming an activity of dissolved $Al = 10^{-4}$. Above pH 11.3, aqueous AlO_2^- occurs. For a lower activity, 10^{-6}, the pH range for $Al(OH)_3$ solid is 4.4–7.25.

Table 18. Thermodynamic data for aluminum

Species (state)	ΔG_f^0 (kcal/gfw)	Reference
Al^{3+} (aq)	−115.92	Wagman et al. (1982)
AlO_2^- (aq)	−198.59	Wagman et al. (1982)
$Al(OH)_3$ (gibbsite)	−276.08	Wagman et al. (1982)
$AlOH^{2+}$ (aq)	−165.89	Wagman et al. (1982)

Abbreviations see Table 1

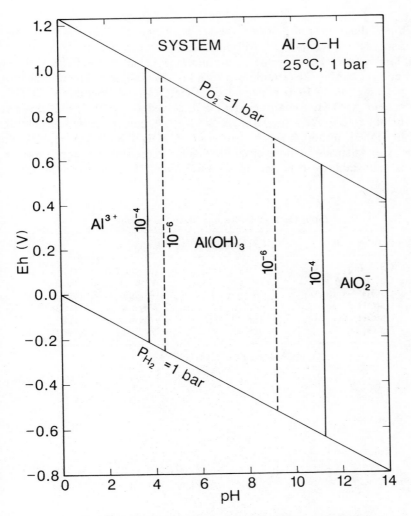

Fig. 20. Eh-pH diagram for part of the system Al-O-H. The assumed activity of dissolved Al = $10^{-4, -6}$. See text for discussion

GALLIUM

The Eh-pH diagram for gallium species is shown in Fig. 21. The thermodynamic data for important gallium species are given in Table 19.

Gallium is strongly diadochic with aluminum in nature, and this is easily seen by comparing Fig. 20 (aluminum) with Fig. 21 (gallium). The middle part of the pH range (5.3–10.4) is occupied by $Ga(OH)_3$, analogous to $Al(OH)_3$ in the system Al-O-H, for an activity of dissolved Ga = 10^{-6}, and this pH range shrinks to 5.9–9.75 for an activity of 10^{-8}. Under acidic pH Ga^{3+} is shown in Fig. 21, although this hydrolyzes to $GaOH^{2+}$ and $Ga(OH)_2^+$, again analogous to Al species. Under basic pH GaO_3^{3-} is the stable aqueous species. Ga_2O_3 is metastable with respect to $Ga(OH)_3$ for 25°C, 1 bar conditions.

Table 19. Thermodynamic data for gallium

Species (state)	ΔG_f^0 (kcal/gfw)	Reference
Ga^{3+} (aq)	−38.00	Wagman et al. (1982)
GaO_3^{3-} (aq)	−147.94	Wagman et al. (1982)
$GaOH^{2+}$ (aq)	−90.89	Wagman et al. (1982)
$Ga(OH)_2^+$ (aq)	−142.78	Wagman et al. (1982)
$Ga(OH)_3$ (c)	−198.69	Wagman et al. (1983)
Ga_2O_3 (c)	−238.53	Wagman et al. (1982)

Abbreviations see Table 1

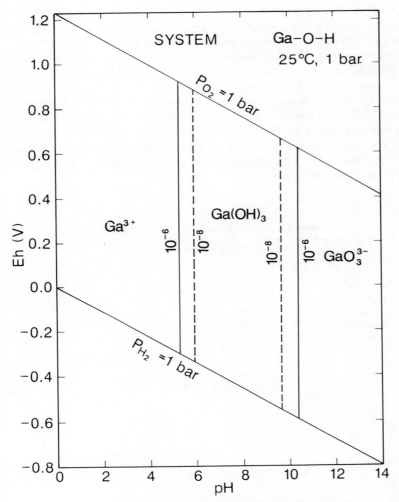

Fig. 21. Eh-pH diagram for part of the system Ga-O-H.
The assumed activity for dissolved Ga = 10^{-8}. See text for discussion

INDIUM

The Eh-pH diagram for indium species is shown in Fig. 22. The thermodynamic data for important indium species are given in Table 20.

Indium is an extremely rare element in the earth's crust. In the system In-O-H-S (Fig. 22), it is noted that below the sulfide-sulfate boundary the species In_2S_3 occupies all the Eh-pH space. Above this boundary, In_2O_3 is stable above pH 5.5 for an assumed activity of dissolved In $= 10^{-8}$. Under acidic conditions, In^{3+} and $InOH^{2+}$ occur, and $In(OH)_2^+$ is metastable with respect to these species. Divalent In(II) may be important under very low oxygen fugacity conditions. InS and In^{2+} are both metastable with respect to In(III) species In_2S_3 and In^{3+}. Under sulfur-absent conditions, In^{2+} will appear in the water stability field.

Table 20. Thermodynamic data for indium

Species (state)	ΔG_f^0 (kcal/gfw)	Reference
In^{2+} (aq)	−12.12	Wagman et al. (1982)
In^{3+} (aq)	−23.42	Wagman et al. (1982)
In_2O_3 (c)	−198.54	Wagman et al. (1982)
$InOH^{2+}$ (aq)	−74.81	Wagman et al. (1982)
$In(OH)_2^+$ (aq)	−125.48	Wagman et al. (1982)
InS (c)	−31.50	Wagman et al. (1982)
In_2S_3 (c)	−98.59	Wagman et al. (1982)
InS (c)	−31.50	Wagman et al. (1982)

Abbreviations see Table 1

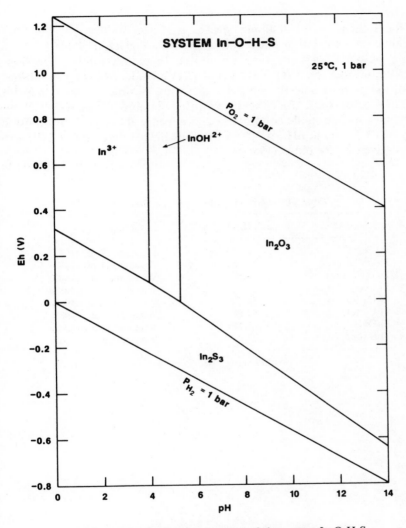

Fig. 22. Eh-pH diagram for part of the system In-O-H-S.
The assumed activities for dissolved species are: In = 10^{-8}, S = 10^{-3}.
See text for discussion

THALLIUM

The Eh-pH diagram for thallium species is shown in Fig. 23. The thermodynamic data for important thallium species are given in Table 21.

The thallium Eh-pH diagram is complex. In nature, thallium possesses three common valences, Tl(I), Tl(III), and Tl(IV). Under oxidizing conditions, thallium oxides dominate the Eh-pH space with, in order of ascending Eh, Tl_2O, Tl_2O_3, and Tl_2O_4. Under acidic oxidizing conditions a field of Tl^{3+} appears. Under sulfide-stable conditions, Tl_2S appears above pH = 7.9 for a Tl activity of 10^{-6}, and at pH for 10^8. A previous Eh-pH diagram for Tl species has been given by Brookins (1986a).

Table 21. Thermodynamic data for thallium

Species (state)	ΔG_f^0 (kcal/gfw)	Reference
Tl^+ (aq)	−7.74	Wagman et al. (1982)
$Tl(OH)_2^+$ (aq)	−58.48	Wagman et al. (1982)
$TlOH^{2+}$ (aq)	−3.80	Wagman et al. (1982)
Tl_2O (c)	−35.20	Wagman et al. (1982)
Tl_2O_3 (c)	−74.50	Wagman et al. (1982)
Tl_2O_4 (c)	−82.98	Wagman et al. (1982)
$Tl(OH)_3$ (c)	−121.18	Wagman et al. (1982)
Tl_2S (c)	−22.39	Wagman et al. (1982)

Abbreviations see Table 1

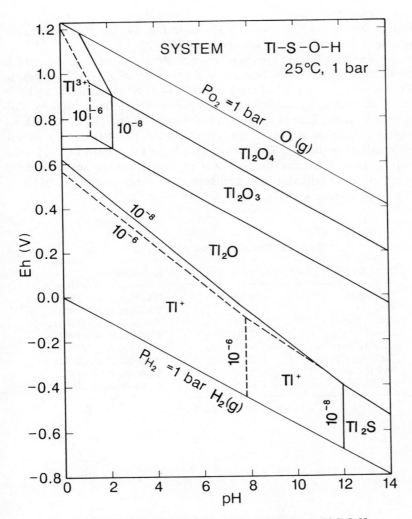

Fig. 23. Eh-pH diagram for part of the system Tl-S-O-H. The assumed activities for dissolved species are: Tl = $10^{-8, -6}$, S = 10^{-3}. See text for discussion

ZINC

The Eh-pH diagram for zinc species is shown in Fig. 24. The thermodynamic data for important zinc species are given in Table 22.

Zinc possesses only one common valence Zn(II) in nature, yet zinc is dependent on redox conditions because of the stability of ZnS (sphalerite). In Fig. 24 the schematics of part of the system Zn-O-H-S-C are shown, and it is noted that under reducing conditions, below the sulfide-sulfate boundary, ZnS occupies much of the Eh-pH space. ZnS dissolves to Zn^{2+} at about pH 2.1 assuming an activity of dissolved zinc of 10^{-6} (pH = 1.14 for 10^{-4}). Above the sulfide-sulfate boundary, Zn^{2+} occupies a large Eh-pH field to pH = 7.5 ($a_{Zn} = 10^{-4}$), then a narrow field of $ZnCO_3$ (smithsonite), followed by a larger field of ZnO (zincite), which dissolves to form ZnO_2^{2-} at pH = 11 ($a_{Zn} = 10^{-8}$) or 13 ($a_{Zn} = 10^{-4}$).

Table 22. Thermodynamic data for zinc

Species (state)	ΔG_f^0 (kcal/gfw)	Reference
Zn^{2+} (aq)	−35.15	Wagman et al. (1982)
ZnO (c)	−76.08	Wagman et al. (1982)
ZnO_2^{2-} (aq)	−91.84	Wagman et al. (1982)
$ZnOH^+$ (aq)	−78.90	Wagman et al. (1982)
$HZnO_2^-$ (aq)	−109.42	Wagman et al. (1982)
$Zn(OH)_2$ (c)	−132.36	Wagman et al. (1982)
ZnS (c)	−48.11	Wagman et al. (1982)
$ZnCO_3$ (c)	−174.84	Wagman et al. (1982)

Abbreviations see Table 1

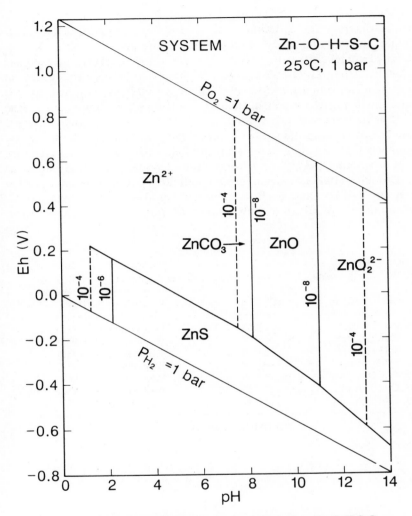

Fig. 24. Eh-pH diagram for part of the system Zn-O-H-S-C. The assumed activities for dissolved species are: $Zn = 10^{-6, -4}$, $C = 10^{-3}$, $S = 10^{-3}$. See text for discussion

CADMIUM

The Eh-pH diagram for cadmium species is shown in Fig. 25. The thermodynamic data for important cadmium species are given in Table 23.

The Eh-pH diagram for the system Cd-C-S-O-H (Fig. 25) is fairly similar to that for the system Zn-O-H-S-C (Fig. 24) which is to be suspected based on their geochemical similarities. In Fig. 25, $a_{Cd} = 10^{-8}$ is assumed for all boundaries involving dissolved Cd. Below the sulfide-sulfate boundary, a large field of CdS (greenockite) exists, although in nature much of the Cd under these conditions is camouflaged by Zn in minerals such as sphalerite. Above this boundary, Cd^{2+} occupies a large Eh-pH field under slightly basic to very acidic pH. At higher pHs, $CdCO_3$ forms, then $Cd(OH)_2$, and then aqueous CdO_2^{2-}. Brookins (1986a) has published a preliminary Eh-pH diagram for Cd species for 25°C, 1 bar; and he (Brookins 1979) has also calculated the 200°C, 1 bar diagram in conjunction with studies of the Oklo Natural Reactor, Gabon.

Table 23. Thermodynamic data for cadmium

Species (state)	ΔG_f^0 (kcal/gfw)	Reference
Cd^{2+} (aq)	−18.55	Wagman et al. (1982)
$Cd(OH)_2$ (c)	−113.19	Wagman et al. (1982)
CdS (c)	−37.40	Wagman et al. (1982)
CdO_2^{2-} (aq)	−67.97	Wagman et al. (1982)
$CdCO_3$ (c)	−160.00	Wagman et al. (1982)
$CdSO_4$ (c)	−196.33	Wagman et al. (1982)
CdO (c)	−54.59	Wagman et al. (1982)

Abbreviations see Table 1

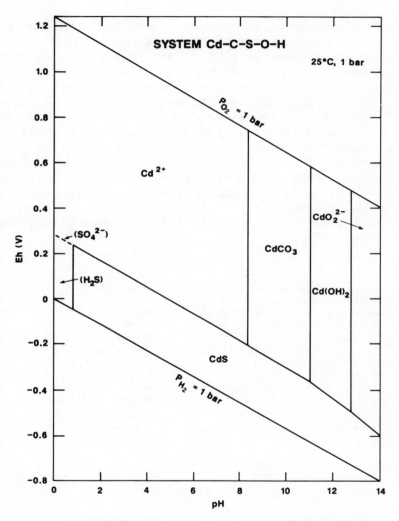

Fig. 25. Eh-pH diagram for part of the system Cd-C-S-O-H.
The assumed activities for dissolved species are: Cd = 10^{-8}, C = 10^{-3}, S = 10^{-3}.
See text for discussion

MERCURY

The Eh-pH diagram for mercury species is shown in Fig. 26. The thermodynamic data for important mercury species are given in Table 24.

Mercury species in the system Hg-S-O-H are shown by solid lines in Fig. 26. Dashed lines are used to show fields for $HgCl_4^{2-}$ and Hg_2Cl_2 (calomel), if Cl is added to the Hg-S-O-H system. In the absence of Cl, the oxidizing, acidic part of the Eh-pH diagram is occupied by mercurous ion Hg_2^{2-} and mercuric ion Hg^{2+}. Most of the Eh-pH diagram above the sulfide-sulfate boundary is occupied by a large field of native Hg. Montroydite (HgO) forms from native mercury at higher redox between pH = 5.4 to 10.6. Montroydite dissolves to form $HHgO_2^-$. For dissolved Hg with an activity of 10^{-9}, the fields of mercurous and mercuric ions are replaced by the chloride species of Hg as shown ($a_{Cl} = 10^{-3.5}$), showing the importance of Cl on mercury transport under oxidizing, acidic conditions.

Previous diagrams for the system Hg-O-H-S-Cl have been given by Parks and Nordström (1979).

Table 24. Thermodynamic data for mercury

Species (state)	ΔG_f^0 (kcal/gfw)	Reference
Hg_2^{2+} (aq)	+36.69	Wagman et al. (1982)
Hg^{2+} (aq)	+39.29	Wagman et al. (1982)
$HgOH^{1+}$ (aq)	−12.50	Wagman et al. (1982)
$HHgO_2^-$ (aq)	−45.48	Wagman et al. (1982)
HgS (c)	−12.09	Wagman et al. (1982)
HgS_2^{2-} (aq)	+41.90	Wagman et al. (1982)
HgO (c)	−58.56	Wagman et al. (1982)
Hg_2Cl_2 (c)	−50.38	Wagman et al. (1982)
$HgCl_4^{2-}$ (aq)	−106.79	Wagman et al. (1982)

Abbreviations see Table 1

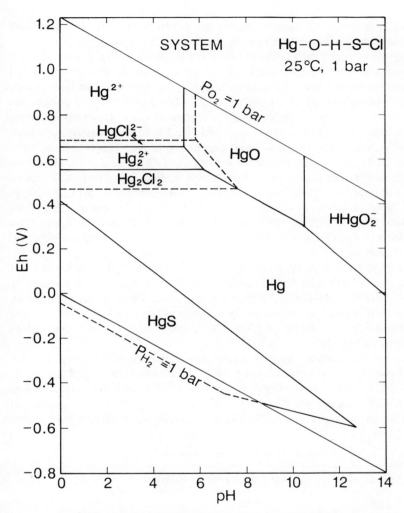

Fig. 26. Eh-pH diagram for part of the system Hg-O-H-S-Cl. The assumed activities for dissolved species are: Hg = 10^{-8}, Cl = $^{-3.5}$, S = 10^{-3}. See text for discussion

COPPER

Three Eh-pH diagrams for copper species are shown as Figs. 27–29. The thermodynamic data for important copper species are given in Table 25.

The simple system Cu-O-H Eh-pH diagram is shown in Fig. 27. An activity of dissolved Cu = 10^{-6} is assumed. The diagram is characterized by a large field of native copper under reducing conditions, but note that this field persits above the hypothetical sulfide-sulfate boundary (see Fig. 1). Native copper oxidizes to cupric ion, Cu^{2+}, under acidic pH, then to Cu_2O (cuprite). Cuprite, in turn, oxidizes to CuO (tenorite) under higher Eh conditions. At very high pH, CuO dissolves to form CuO_2^-, and cuprite oxidizes to this species as well.

In Fig. 28 the phase relations in the system Cu-S-O-H are shown. Here, two important sulfides appear below the sulfide-sulfate boundary. Chalcocite (Cu_2S) and covellite (CuS) both occupy significant parts of the Eh-pH diagram. Note, however, that above the sulfide-sulfate boundary, as sulfate forms from S(–II), Cu^{1+} from chalcocite is reduced to native copper. The relative fields of Cu_2O, CuO, CuO_2^{2-}, and most of the Cu^{2+} field are identical to those shown in Fig. 27.

Figure 29 shows the phase relations when C is added. Malachite is the most important cupric carbonate in the system Cu-C-S-O-H, as azurite is metastable to it. Malachite ($Cu_2(OH)_2CO_3$) replaces the tenorite fields (Figs. 27 and 28), and forms from cuprite as indicated. Two contours for $a_c = 10^{-1}$ and 10^{-3} are shown in Fig. 29. The fields of CuS, Cu_2S, and native Cu are identical to those shown in Fig. 28.

Numerous diagrams for copper species have been published previously, including those by Garrels and Christ (1965), Anderson (1982), and others. If Fe is added to the Cu-C-S-O-H system, additional complexities arise, including phases such as chalcopyrite ($CuFeS_2$) and bornite (Cu_5FeS_4), and two Eh-pH diagrams for the system Cu-Fe-C-S-O-H are given in Garrels and Christ (1965; pp. 231, 232).

Table 25. Thermodynamic data for copper

Species (state)	ΔG_f^0 (kcal/gfw)	Reference
Cu (c)	0.00	Wagman et al. (1982)
Cu^+ (aq)	+11.94	Wagman et al. (1982)
Cu^{2+} (aq)	+15.65	Wagman et al. (1982)
Cu_2S (c)	−20.60	Wagman et al. (1982)
CuS (c)	−12.81	Wagman et al. (1982)
Cu_2O (c)	−34.98	Garrels and Christ (1965)
CuO (c)	−31.00	Wagman et al. (1982)
CuO_2^{2-} (aq)	−43.88	Wagman et al. (1982)
$Cu_2(CO_3)(OH)_2$ (c)	−213.58	Wagman et al. (1982)
$Cu_3(CO_3)_2(OH)_2$ (c)	−314.29	Wagman et al. (1982)

Abbreviations see Table 1

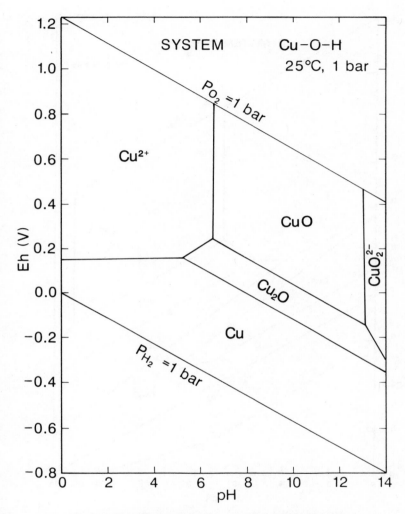

Fig. 27. Eh-pH diagram for part of the system Cu-O-H. The assumed activity for dissolved Cu = 10^{-6}. See text for discussion

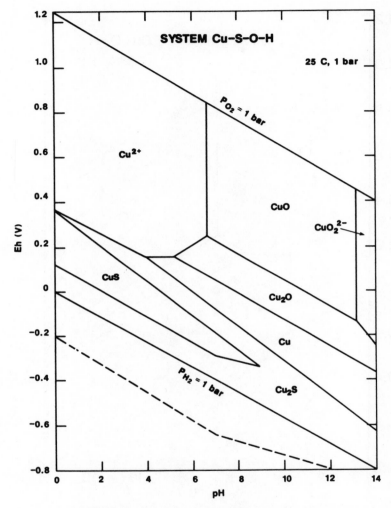

Fig. 28. Eh-pH diagram for part of the system Cu-S-O-H. The assumed activities for dissolved species are: $Cu = 10^{-6}$, $S = 10^{-3}$. See text for discussion

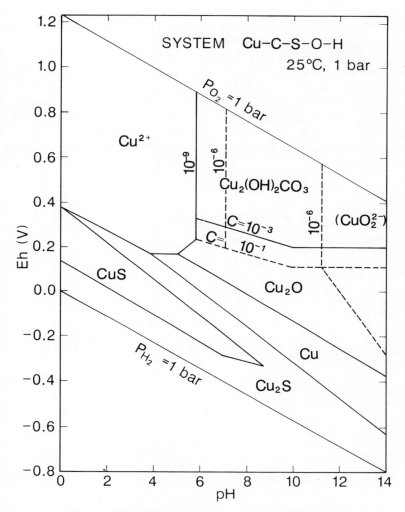

Fig. 29. Eh-pH diagram for part of the system Cu-C-S-O-H. The assumed activities for dissolved species are: $Cu = 10^{-6}$, $S = 10^{-3}$, $C = 10^{-1,-3}$. See text for discussion

SILVER

The Eh-pH diagram for silver species is shown in Fig. 30. The thermodynamic data for important silver species are given in Table 26.

The system Ag-S-O-H in Eh-pH space is dominated by a very large field of native silver. Under reducing conditions, Ag_2S (argenite) occupies a major part of the Eh-pH space as well. Note in Fig. 30 that native Ag can coexist with S(−II) and S(VI) aqueous species as a function of Eh. At high Eh, Ag oxidizes to Ag^+ or $Ag(OH)_2^-$ ($a_{Ag} = 10^{-8}$). When Cl is added to the system Ag-S-O-H, a field of $AgCl_2^-$ replaces much of the field for Ag^+ ($a_{Cl} = 10^{-3.5}$, showing the importance of dissolved Cl on Ag transport under oxidizing, acidic conditions.

Table 26. Thermodynamic data for silver

Species (state)	ΔG_f^0 (kcal/gfw)	Reference
Ag^{2+} (aq)	+64.29	Wagman et al. (1982)
Ag^+ (aq)	−18.43	Wagman et al. (1982)
Ag_2O (c)	−2.68	Wagman et al. (1982)
Ag_2O_2 (c)	+6.60	Wagman et al. (1982)
Ag_2O_3 (c)	+29.01	Wagman et al. (1982)
$Ag(OH)_2^-$ (aq)	−62.19	Wagman et al. (1982)
Ag_2S (c)	−9.72	Wagman et al. (1982)
$AgCl_2^-$ (aq)	−51.48	Wagman et al. (1982)

Abbreviations see Table 1

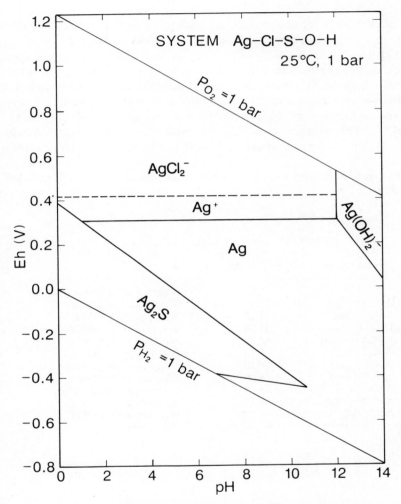

Fig. 30. Eh-pH diagram for part of the system Ag-Cl-S-O-H.
The assumed activities for dissolved species are: Ag = 10^{-8}, S = 10^{-3}, Cl = $10^{-3.5}$.
See text for discussion

GOLD

The Eh-pH diagrams for gold species are shown in Figs. 31 and 32. The thermodynamic data for important gold species are given in Table 27.

Gold species are shown in the Eh-pH space in Fig. 31, and, as expected, native gold occupies essentially all of the Eh-pH space. Yet this diagram, which intentionally includes dissolved Cl as well, illustrates that some gold transport is possible under very acidic, most oxidizing conditions ($a_{Cl} = 10^{-3.5}$). Further, increasing the activity of dissolved Cl will enlarge the fields for $AuCl_2^{4-}$ and $AuCl_2^-$. This is shown in Fig. 32, where $a_{Cl} = 10^{-1}$ coupled with $a_{Au} = 10^{-12}$ shows a large amount of Eh-pH occupied by Au-chloride complexes.

Further, gold transport may also occur by sulfide complexes (Garrels and Christ 1965).

Table 27. Thermodynamic data for gold

Species (state)	ΔG_f^0 (kcal/gfw)	Reference
Au^0 (c)	0.00	Wagman et al. (1982)
$Au(OH)_3$ (c)	−75.75	Wagman et al. (1982)
AuO_3^{3-} (aq)	−12.38	Wagman et al. (1982)
$HAuO_3^{2-}$ (aq)	−33.99	Wagman et al. (1982)
$H_2AuO_3^{1-}$ (aq)	−52.17	Wagman et al. (1982)
$AuCl_2^{1-}$ (aq)	−36.12	Wagman et al. (1982)
$AuCl_4^{1-}$ (aq)	−56.20	Wagman et al. (1982)
$HAuCl_4$ (aq)	−56.20	Wagman et al. (1982)
AuO_2 (c)	+48.00	Garrels and Christ (1965)
$Au(CN)_2^{1-}$ (aq)	−68.31	Wagman et al. (1982)
$Au(SCN)_2^{1-}$ (aq)	−60.21	Wagman et al. (1982)

Abbreviations see Table 1

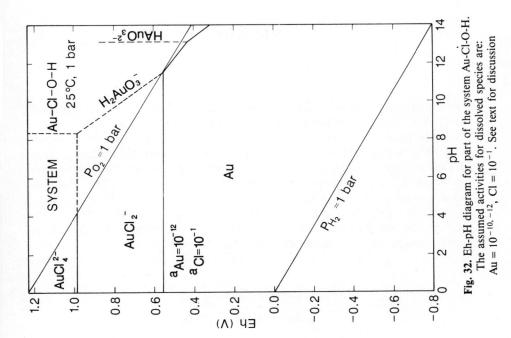

Fig. 32. Eh-pH diagram for part of the system Au-Cl-O-H. The assumed activities for dissolved species are: Au = $10^{-10, -12}$, Cl = 10^{-1}. See text for discussion

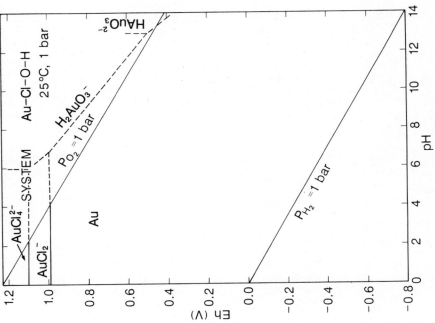

Fig. 31. Eh-pH diagram for part of the system Au-Cl-O-H. The assumed activities for dissolved species are: Au = $10^{-8, -10}$, Cl = $10^{-3.5}$. See text for discussion

NICKEL

The Eh-pH diagrams for nickel species are shown in Figs. 33 and 34. The thermodynamic data for important nickel species are given in Table 28.

In the system Ni-O-H and for $a_{Ni} = 10^{-4}$ or 10^{-6} (Fig. 33), a fairly narrow field of $Ni(OH)_2$ (pH 8–12 or 9–11) separates a large field of Ni^{2+} from $HNiO_2^-$. Thus, nickel mobility in sulfur-poor systems under near-neutral to acidic conditions is obvious. The importance of S is shown in Fig. 34, where a large field of millerite (NiS) occurs below the sulfide-sulfate boundary ($a_S = 10^{-3}$). Not shown in Figs. 33 and 34 are the species $NiOH^+$ and $NiCO_3$, as both are metastable to Ni^{2+} and $Ni(OH)_2$, respectively. Unlike cobalt, nickel does not form a stable carbonate species.

Table 28. Thermodynamic data for nickel

Species (state)	ΔG_f^0 (kcal/gfw)	Reference
Ni^{2+} (aq)	−10.90	Wagman et al. (1982)
NiO (c)	−50.60	Wagman et al. (1982)
$Ni(OH)_2$ (c)	−106.88	Wagman et al. (1982)
NiS (c)	−19.00	Wagman et al. (1982)
$NiSO_4$ (c)	−181.57	Wagman et al. (1982)
$NiCO_3$ (c)	−146.39	Wagman et al. (1982)
$NiOH^+$ (c)	−54.40	Wagman et al. (1982)
$HNiO_2^-$ (aq)	−83.46	Garrels and Christ (1965)

Abbreviations see Table 1

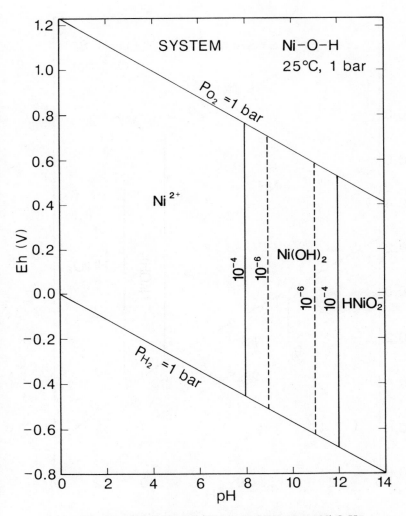

Fig. 33. Eh-pH diagram for part of the system Ni-O-H. Assumed activity of dissolved Ni = $10^{-4, -6}$. See text for discussion

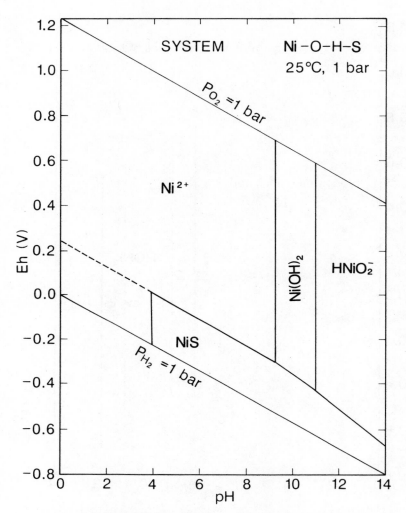

Fig. 34. Eh-pH diagram for part of the system Ni-O-H-S. Assumed activities for dissolved species are: Ni = $10^{-4, -6}$, S = 10^{-3}. See text for discussion

COBALT

The Eh-pH diagram for cobalt species is shown in Fig. 35. The thermodynamic data for important cobalt species are given in Table 29.

Phase relations in the system Co-S-C-O-H are shown in Fig. 35 for the following assumed activities: $a_{Co} = 10^{-6}$, $a_C = 10^{-3}$, $a_S = 10^{-3}$. Cobalt and nickel are often cited as a geochemical pair exhibiting near-ideal substitution for each other. This is indeed apparent in igneous rocks and in high temperature magmatic sulfides and in many metamorphic rocks. In the near-to-surface environment, however, differences are apparent as illustrated by comparing Fig. 35 for Co with Figs. 33 and 34 for Ni species. Cobalt possesses a Co(III) valence, and a large field of Co_3O_4 occurs under oxidizing, neutral to basic pH (Fig. 35). In addition, cobalt forms a stable carbonate (spherocobaltite; $CoCO_3$), whereas nickel does not (cf. Figs. 33 and 34 with 35). Using old thermodynamic data, Garrels and Christ (1965) have reported Eh-pH diagrams for Ni and Co species showing a stable field for Ni_3O_4 and $Co(OH)_3$, respectively. The data in Tables 28 (Ni) and 29 (Co) are better, however, and these species do not appear in the water stability field. $Co(OH)_3$ oxidizing from Co_3O_4 yields Eh = 1.76-0.059 pH, for example, well above the upper stability limit of water. Data for Ni_3O_4 (Garrels and Christ 1965; see Fig. 7.29a, b, pp. 244, 245) are not given. The boundary between $Ni(OH)_2$ and $Ni(OH)_3$, however, yields Eh = 1.5-0.59 pH, again well above the upper stability limit of water.

Table 29. Thermodynamic data for cobalt

Species (state)	ΔG_f^0 (kcal/gfw)	Reference
Co^{2+} (aq)	−13.00	Wagman et al. (1982)
Co^{3+} (aq)	+32.03	Wagman et al. (1982)
CoO (c)	−51.20	Wagman et al. (1982)
Co_3O_4 (c)	−184.99	Wagman et al. (1982)
$HCoO_2^-$ (aq)	−82.97	Garrels and Christ (1965)
$Co(OH)_2$ (c)	−107.53	Wagman et al. (1982)
CoS (c)	−19.80	Garrels and Christ (1965)
$CoCO_3$ (c)	−155.57	Garrels and Christ (1965)
$Co(OH)_3$ (c)	−142.60	Garrels and Christ (1965)

Abbreviations see Table 1

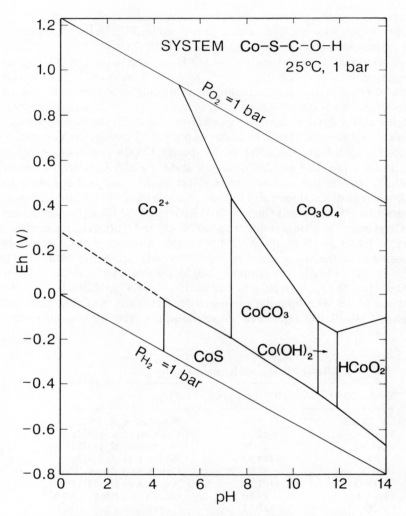

Fig. 35. Eh-pH diagram for part of the system Co-S-C-O-H.
Assumed activities for dissolved species are: Co = 10^{-6}, C = 10^{-3}, S = 10^{-3}.
See text for discussion

IRON

The Eh-pH diagrams for various iron species are given in Figs. 36 through 42. The thermodynamic data for important iron species are given in Table 30.

Several Eh-pH diagrams for the system Fe-($\pm$C)-($\pm$S)-O-H are presented here. Figure 36 shows the phase relations in the simple system Fe-O-H assuming $Fe(OH)_3$ is the stable phase for the Fe(III) precipitate. Further, only Fe^{3+} is considered of the many Fe(III) aqueous species [$FeOH^{2+}$, $Fe(OH)_2^+$, $Fe_2(OH)_4^{2+}$, etc.], but this is justified since the total field of aqueous Fe(III) species is in the small Eh-pH field shown in Fig. 36. The assumption $a_{Fe} = 10^{-6}$ is also used. This figure is characterized by a large field of $Fe(OH)_3$, a large field of Fe^{2+}, and under reducing, basic conditions, a field of $Fe(OH)_2$. Magnetite is not considered in this diagram.

Addition of C to the simple Fe-O-H system is shown in Fig. 37. Here, a small field of $FeCO_3$ appears in the pH range of 6.8 to 9.4, separating Fe^{2+} from $Fe(OH)_2$, and encroaching slightly on the $Fe(OH)_3$ field. Again, magnetite is not assumed in this diagram.

The effect of adding Si to the system Fe-O-H, as the field of $Fe(OH)_2$ is replaced by $FeSiO_3$, is shown in Fig. 38. $Fe(OH)_3$ is again considered as the stable Fe(III) phase, and magnetite is again assumed to be absent.

Figure 39 shows the phase relations in the system Fe-Si-O-H, now considering magnetite (Fe II, III) as well as Fe(II) and Fe(III) phases. With $Fe(OH)_3$ again chosen as the stable Fe(III) phase, a very large magnetite field appears. Klein and Bricker (1977) have used a similar diagram to comment on the origin of banded iron formations. The $FeSiO_3$ field replaces the $Fe(OH)_2$ field of Fig. 36. To a first approximation, the $FeSiO_3$ field can be assumed to be close to that for Fe(II) silicates found in banded iron formations.

Another version of the Fe-O-H-Si system (Fig. 40) considers $FeO \cdot OH$ (goethite) as the stable Fe(III) phase. Magnetite still exists, separating $FeSiO_3$ from goethite, but its Eh-pH stability field is much smaller than when $Fe(OH)_3$ is considered (Fig. 39).

Addition of C to the Fe-Si-O-H system with $FeO \cdot OH$ as the stable Fe(III) phase is shown in Fig. 41, with, similar to Fig. 37, the siderite field separating Fe^{2+} from both Fe_3O_4 and $FeSiO_3$, and encroaching slightly into the $FeO \cdot OH$ field.

Finally, the effect of sulfur on the Fe-O-H system is considered in Fig. 42. Here, the reducing conditions below the sulfide-sulfate boundary yield a large field of pyrite which, along with the very small pyrrhotite field (at highest pH and most reducing conditions), obliterates most of the magnetite field and all the $FeSiO_3$ [or $Fe(OH)_2$] field, and also the field of $FeCO_3$. Hematite (Fe_2O_3) is shown as the stable Fe(III) phase, since both $Fe(OH)_3$ and $FeO \cdot OH$ will age to Fe_2O_3 (although the kinetics for this aging may be very slow). Figure 42 is the standard reference diagram for Fe species. Garrels and Christ (1965) and Drever (1982) have presented numerous diagrams for Fe systematics, and the reader is referred to their work for more detail. They point out, for example, that even when S activity is greatly reduced (to near 10^{-6}), a small field of pyrite still exists in the neutral pH range along the sulfide-sulfate boundary, demonstrating its widespread occurrence over widely differing chemical conditions.

Other Fe Eh-pH diagrams have been given by Hem (1977) and more recently by Winters and Buckley (1986) who advocate consideration of aqueous $FeSi_3O_3(OH)_8^0$ as an important species.

Table 30. Thermodynamic data for iron

Species (state)	ΔG_f^0 (kcal/gfw)	Reference
Fe^{2+} (aq)	-18.86	Wagman et al. (1982)
Fe^{3+} (aq)	-1.12	Wagman et al. (1982)
FeO_2^{2-} (aq)	-70.58	Wagman et al. (1982)
Fe_2O_3 (c)	-177.39	Wagman et al. (1982)
Fe_3O_4 (c)	-242.69	Wagman et al. (1982)
$Fe(OH)_3$ (c)	-166.47	Wagman et al. (1982)
$Fe(OH)_2$ (c)	-116.30	Winters and Buckley (1986)
$FeO \cdot OH$ (c)	-116.77	Robie et al. (1978)
FeS_2 (c)	-39.89	Wagman et al. (1982)
FeS (c)	-24.00	Wagman et al. (1982)
$FeSiO_3$ (c)	-267.16	Winters and Buckley (1986)
$FeCO_3$ (c)	-159.34	Winters and Buckley (1986)
$FeSi_3O_3(OH)_8^0$ (aq)	-898.00	Winters and Buckley (1986)

Abbreviations see Table 1

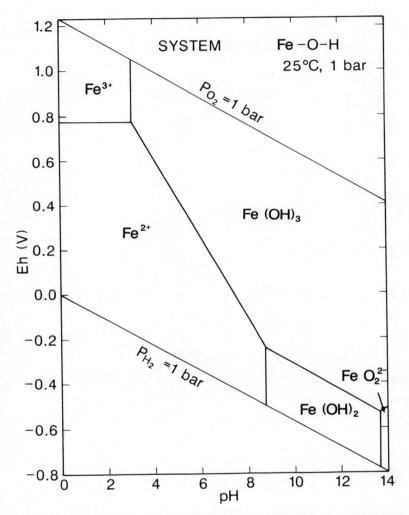

Fig. 36. Eh-pH diagram for part of the system Fe-O-H assuming Fe(OH)$_3$ as stable Fe(III) phase. Assumed activity of dissolved Fe = 10^{-6}. See text for discussion

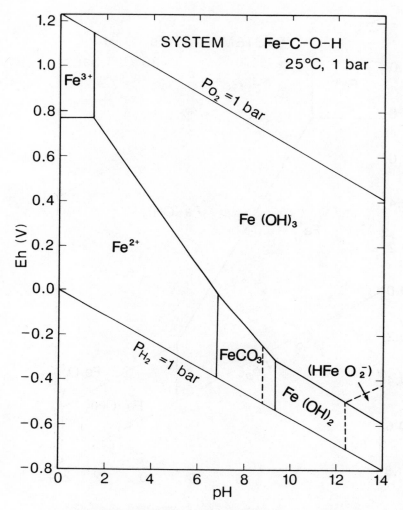

Fig. 37. Eh-pH diagram for part of the system Fe-C-O-H. Assumed activities of dissolved species are: Fe = 10^{-6}, C = 10^{-3}. See text for discussion

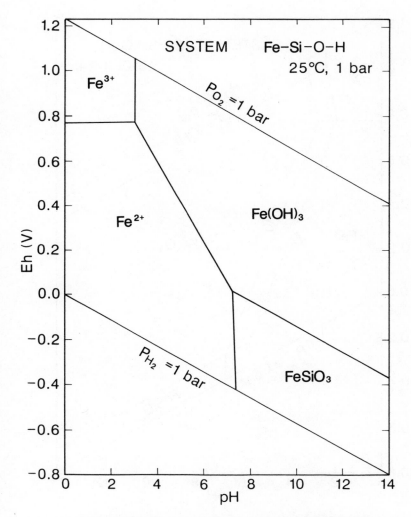

Fig. 38. Eh-pH diagram for part of the system Fe-Si-O-H. Assumed activities for dissolved species are: Fe = 10^{-6}, Si = 10^{-3}. Fe(OH)$_3$ is assumed as the dominant Fe(III) phase and magnetite is assumed to be absent. See text for discussion

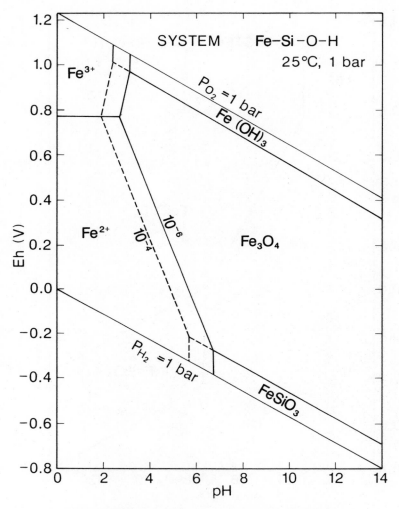

Fig. 39. Eh-pH diagram for part of the system Fe-Si-O-H.
Assumed activities for dissolved species are: Fe = 10^{-6}, Si = 10^{-3}. Fe(III) phases assumed to be Fe(OH)$_3$ and magnetite. See text for discussion

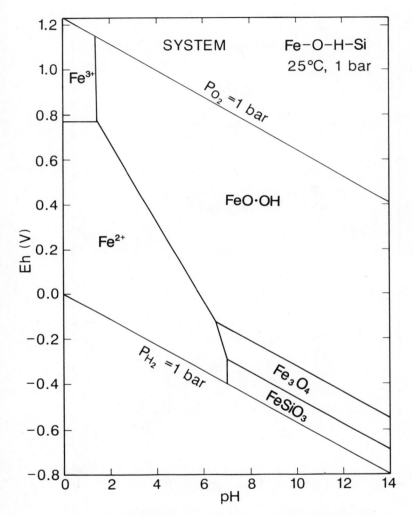

Fig. 40. Eh-pH diagram for part of the system Fe-O-H-Si. Assumed activities for dissolved species are: Fe = 10^{-6}, Si = 10^{-3}. Fe(III) assumed to be present in both goethite and magnetite. See text for discussion

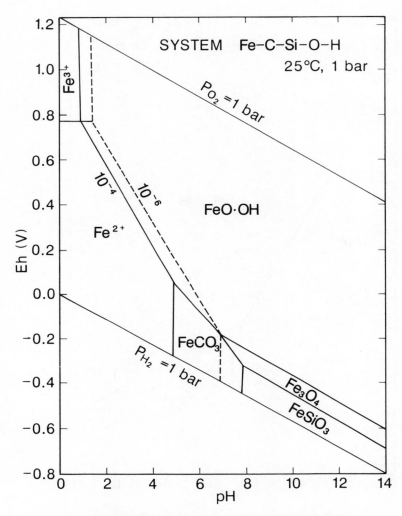

Fig. 41. Eh-pH diagram for part of the system Fe-C-Si-O-H. Assumed activities for dissolved species are: Fe = 10^{-6}, Si = 10^{-3}, C = 10^{-3}. Goethite and magnetite assumed as Fe(III) solid phases. See text for discussion

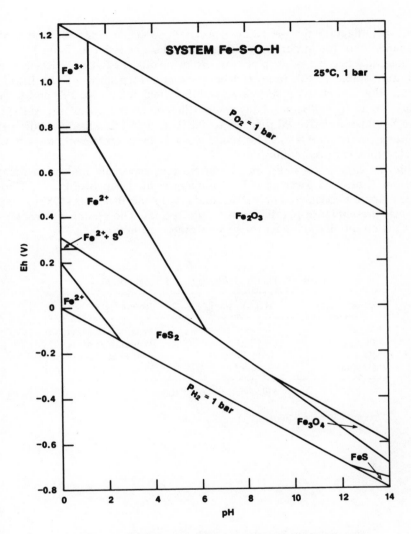

Fig. 42. Eh-pH diagram for part of the system Fe-S-O-H.
Assumed activities of dissolved species are: Fe = 10^{-6}, S = 10^{-3}. Hematite is assumed to be the major Fe(III) phase. See text for discussion

PALLADIUM

The Eh-pH diagram for palladium species is shown in Fig. 43. The thermodynamic data for important palladium species are given in Table 31.

Palladium, like the other platinum group metals (Pt-Ir-Os-Pd-Rh-Ru), is remarkably stable in the terrestrial environment. Inspection of the Eh-pH diagram (Fig. 43) shows a very large stability field for native Pd, and large stability fields for PdS and PdS_2 under Eh conditions below the sulfide-sulfate boundary. Native Pd oxidizes to Pd(II) in Pd^{2+} (oxidizing, acidic pH conditions) or to $Pd(OH)_2$. The phase PdO is metastable with respect to $Pd(OH)_2$ under normal conditions.

Chloride complexes of Pd may be important, especially under sea water conditions (Baes and Mesmer 1976; Goldberg et al. 1986; Brookins 1987b), but these are not considered here. Westland (1981) and Brookins (1987b) have previously presented Eh-pH diagrams for Pd species. The version of Brookins (1987b) is essentially identical to the one presented here.

Table 31. Thermodynamic data for palladium

Species (state)	ΔG_f^0 (kcal/gfw)	Reference
PdS (c)	−16.01	Wagman et al. (1982)
PdS_2 (c)	−17.81	Wagman et al. (1982)
PdO (c)	−15.31	Pourbaix (1966)
Pd^{2+} (aq)	+42.49	Wagman et al. (1982)

Abbreviations see Table 1

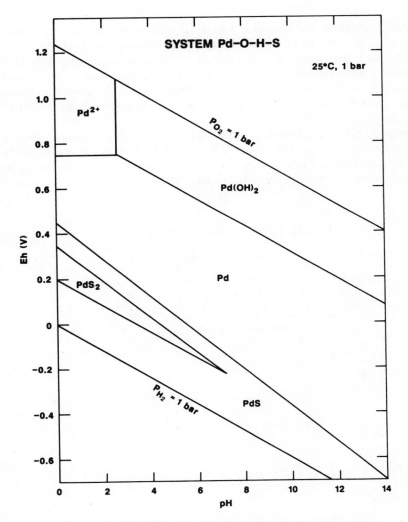

Fig. 43. Eh-pH diagram for part of the system Pd-O-H-S. Assumed activities of dissolved species are: $Pd = 10^{-8}$, $S = 10^{-3}$. See text for discussion

RHODIUM

The Eh-pH diagram for rhodium species is shown in Fig. 44. The thermodynamic data for important rhodium species are given in Table 32.

Unlike the other platinum group metals, rhodium is unique in that it does not form any sulfide minerals. Instead, native Rh occupies most of the Eh-pH space shown in Fig. 44, both above and below the sulfide-sulfate fence. Rhodium of Rh(I) and Rh(III) valences are known. Stable fields of Rh_2O, Rh^{3+}, and Rh_2O_3 are plotted in Fig. 44.

Westland (1981) and Brookins (1987b) have previously presented Eh-pH diagrams for Rh species. The diagram presented here (Fig. 44) is essentially identical to that presented by Brookins (1987b), who had the advantage to work with more recent data than Westland (1981).

Table 32. Thermodynamic data for rhodium

Species (state)	ΔG_f^0 (kcal/gfw)	Reference
Rh_2O (c)	−20.0	Pourbaix (1966)
Rh_2O_3 (c)	−50.0	Pourbaix (1966)
Rh^{3+} (aq)	+55.3	Pourbaix (1966)

Abbreviations see Table 1

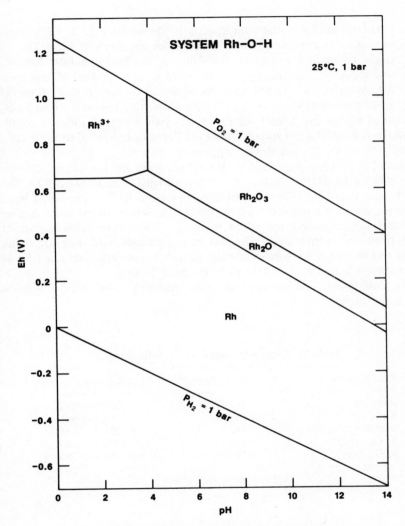

Fig. 44. Eh-pH diagram for part of the system Rh-O-H. Assumed activity for dissolved Rh = $^{-8}$. See text for discussion

RUTHENIUM

The Eh-pH diagram for ruthenium species is shown in Fig. 45. The thermodynamic data for important ruthenium species are given in Table 33.

In nature, ruthenium possesses the following valences: Ru(0), Ru(IV), Ru(VI), and Ru(VII). Assuming $a_{Ru} = 10^{-8}$ and $a_S = 10^{-3}$, Fig. 45 has been constructed. Most of the Eh-pH space is occupied by solids: RuS_2 (laurite) under the sulfide-sulfate boundary, native Ru above this boundary, and RuO_2 above that at higher Eh. Under mildly to very oxidizing and low pH conditions, aqueous $Ru(OH)_2^{2+}$ is stable. Under oxidizing near-neutral to basic conditions, fields of RuO_4^- and RuO_4^{2-} appear as shown.

Westland (1981) has presented an Eh-pH diagram for Ru species, but includes fields for Ru(III) oxide $Ru_2O_3 \cdot xH_SO$ and Ru(IV) as $RuO_2 \cdot 2H_2O$. The diagram from Brookins (1987b), which is very close to that presented here, relies on the recently compiled OECD (1985) data which are of much higher quality than data available just a few years ago. Ruthenium is an important element from radioactive waste disposal considerations, and data are being gathered and revised in a rigorous fashion such that it is likely that this Eh-pH diagram shown here may be changed in the near future.

Ruthenium shows that, under some very oxidizing conditions, it may be mobile as an oxyion.

Table 33. Thermodynamic data for ruthenium

Species (state)	ΔG_f^0 (kcal/gfw)	Reference
RuS_2 (c)	−45.03	OECD (1985)
RuO_2 (c)	−60.49	OECD (1985)
$Ru(OH)_2^{2+}$ (aq)	−53.01	OECD (1985)
RuO_4^{2-} (aq)	−73.28	OECD (1985)
RuO_4^- (aq)	−59.78	OECD (1985)

Abbreviations see Table 1

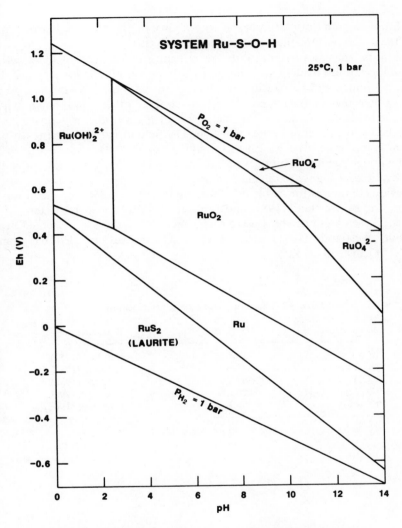

Fig. 45. Eh-pH diagram for part of the system Ru-S-O-H. Assumed activities of dissolved species are: $Ru = 10^{-8}$, $S = 10^{-3}$. See text for discussion

PLATINUM

The Eh-pH diagram for platinum species is shown in Fig. 46. The thermodynamic data for important platinum species are given in Table 34.

Platinum species present in the terrestrial environment include native platinum, platinum oxide-hydroxide, and platinum sulfides. Under the sulfide-sulfate boundary, PtS (cooperite) and PtS_2 occur as shown. As in the case of several other metals (i.e., mercury, silver, copper, other platinoids), when sulfide is oxidized to sulfate, the metal ion in the sulfides is reduced as a counterbalancing attempt to the oxidation. The field of native Pt in the Eh-pH diagram is immense. At high Eh Pt(0) oxidizes to $Pt(OH)_2$ and then to PtO_2 near the upper stability limit of water. A small field of Pt^{2+} occures under extreme acidic and highly oxidizing conditions.

Previous diagrams for Pt include those by Pourbaix (1966) for metal-water systems, and by Westland (1981) and Brookins (1987b) for metal-water-sulfur systems. These diagrams are close to each other and to the one presented here, although Westland (1981) uses $PtO \cdot xH_2O$ instead of $Pt(OH)_2$ and $PtO_2 \cdot xH_2O$ instead of PtO_2.

Table 34. Thermodynamic data for platinum

Species (state)	ΔG_f^0 (kcal/gfw)	Reference
PtS (c)	−21.6	Latimer (1952)
PtS_2 (c)	−25.6	Latimer (1952)
$Pt(OH)_2$ (c)	−68.2	Latimer (1952)
PtO_2 (c)	−20.0	Pourbaix (1966)
Pt^{2+} (aq)	+54.8	Pourbaix (1966)

Abbreviations see Table 1

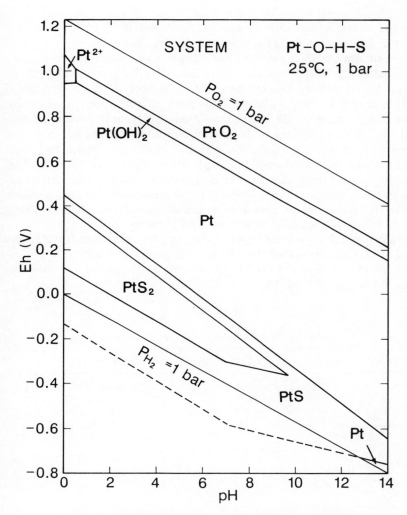

Fig. 46. Eh-pH diagram for part of the system Pt-O-H-S. Assumed activities for dissolved species are: Pt = 10^{-8}, S = 10^{-3}. See text for discussion

IRIDIUM

The Eh-pH diagram for iridium species is shown in Fig 47. The thermodynamic data for important species are given in Table 35.

Native iridium occupies much of the Eh-pH space as shown in Fig. 47. However, below the Eh-pH sulfide-sulfate boundary both IrS_2 and Ir_2S_3 are important. Native iridium also oxidizes to IrO_2 at fairly high Eh, and Ir(IV) to Ir(VI) at higher Eh and basic pH. This field of IrO_4^{2-} may be of importance in nature, as it allows for some segregation of Ir from Pt during weathering, and thus during recycling of sediments. Brookins (1987b) has discussed the importance of this in conjunction with K-T boundary iridium and other platinoid abundances. The dashed line of the IrO_4^{2-} field is for $a_{Ir} = 10^{-10}$ instead of 10^{-8} (solid line). Goldberg et al. (1986) have argued for Ir(III) oxyspecies as being important in sea water. Unfortunately, the thermodynamic data for Ir(III) oxyspecies are poor at best, and all attempts at calculation yield metastable Ir(III) species with respect to Ir(IV) species.

Pourbaix (1966) has presented Eh-pH diagrams for the system Ir-water, and Westland (1981) and Brookins (1987b) for the system Ir-S-water.

Westland (1981), using $IrO_2 \cdot 2H_2O$ instead of IrO_2, did not consider IrO_4^{2-} as did Brookins (1987b).

Table 35. Thermodynamic data for iridium

Species (state)	ΔG_f^0 (kcal/gfw)	Reference
Ir_2S_3 (c)	−49.8	Latimer (1952)
IrS_2 (c)	−30.4	Latimer (1952)
IrO_2 (c)	−28.0	Pourbaix (1966)
IrO_4^- (aq)	−47.0	Pourbaix (1966)

Abbreviations see Table 1

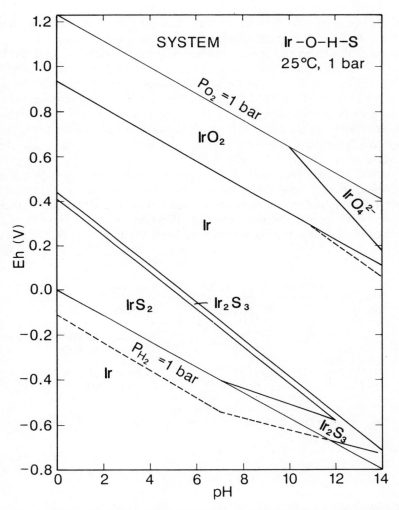

Fig. 47. Eh-pH diagram for part of the system Ir-O-H-S. Assumed activities of dissolved species are: Ir = $10^{-8,-10}$, S = 10^{-3}.
See text for discussion

OSMIUM

The Eh-pH diagram for osmium species is shown in Fig. 48. The thermodynamic data for important osmium species are given in Table 36.

Osmium is similar to ruthenium in many respects, as each form an M(IV) sulfide. The OsS_2 (erlichmanite) is stable below the sulfide-sulfate boundary. Native Os oxidizes to Os(VI) in OsS_2 and is then formed by reduction as S(−II) in OsS_2 is oxidized to S(VI) in sulfate ion. Osmium also possesses Os(VI) and Os(VIII) valences, which form the oxyions OsO_4^{2-} and either H_2OsO_5 or $HOsO_5^-$ as shown in Fig. 48. The potential for Os mobility under oxidizing conditions is thus greater than for any other of the platinoids. Activities assumed are: $a_{Os} = 10^{-8}$, $a_S = 10^{-3}$.

Pourbaix (1966) has presented the Eh-pH diagram for the system Os-water, and Westland (1981) and Brookins (1987b) for the system Os-S-water. Brookins (1987b) had the advantage of more recent and complete thermodynamic data for Os species, hence his diagram, which is close to that presented here, is different in some respects from Westland's (1981).

Table 36. Thermodynamic data for osmium

Species (state)	ΔG_f^0 (kcal/gfw)	Reference
OsS_2 (c)	−27.49	Wagman et al. (1982)[a]
$Os(OH)_4$ (c)	−160.97	Wagman et al. (1982)
H_2OsO_5 (aq)	−128.81	Wagman et al. (1982)
$HOsO_5^-$ (aq)	−112.38	Wagman et al. (1982)
OsO_4^{2-} (aq)	−89.25	Pourbaix (1966)

Abbreviations see Table 1

[a] G_f^0 value for OsS_2 calculated from other (H^0, S^0) data in reference

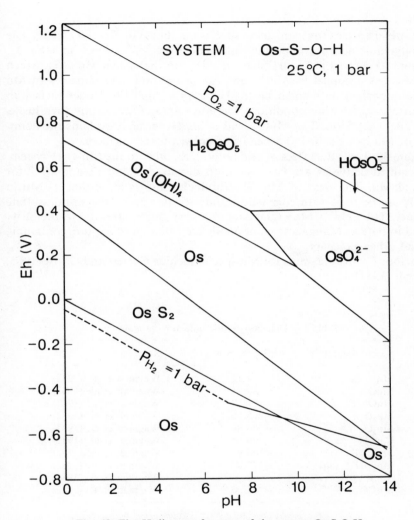

Fig. 48. Eh-pH diagram for part of the system Os-S-O-H. Assumed activities of dissolved species are: Os = 10^{-8}, S = 10^{-3}. See text for discussion

MANGANESE

The Eh-pH diagrams for manganese species are shown in Figs. 49 and 50. The thermodynamic data for important manganese species are given in Table 37.

Figure 49 shows the Eh-pH phase relations in the system Mn-O-H. Much of the field is occupied by Mn^{2+}, and the oxides and oxyhydroxides of Mn become important only under basic pH except at high Eh. Under surface to near-surface weathering conditions, various MnO_2 polymorphs (pyrolusite, todorokite, etc.) form. The MnO_2 field is important as it explains the abundance of Mn-oxide stains and dendrites on weathered surfaces.

Figure 50 shows the effect of adding both S and C to the Mn-O-H system. The assumed activities are 10^{-3} for both dissolved S and C, and 10^{-6} for dissolved Mn. The field of $MnCO_3$ (rhodochrosite) is important as Mn in $MnCO_3$ can be fixed with cogenetic sulfides or just above the sulfide-sulfate boundary. The field of MnS (alabandite) is very small, attesting to the rarity of MnS in nature. Manganese compounds are often quite impure, containing iron and other elements.

Previous Eh-pH diagrams for Mn species include Garrels and Christ (1965) and Hem (1981).

Table 37. Thermodynamic data for manganese

Species (state)	ΔG_f^0 (kcal/gfw)	Reference
Mn^{2+} (aq)	−54.52	Wagman et al. (1982)
MnO (c)	−86.74	Wagman et al. (1982)
MnO_2 (c)	−111.17	Wagman et al. (1982)
Mn_2O_3 (c)	−210.59	Wagman et al. (1982)
Mn_3O_4 (c)	−306.69	Wagman et al. (1982)
$MnOH^+$ (aq)	−96.80	Wagman et al. (1982)
$Mn(OH)_2$ (c)	−146.99	Wagman et al. (1982)
$Mn(OH)_3^-$ (aq)	−177.87	Wagman et al. (1982)
MnS (c)	−52.20	Wagman et al. (1982)
$MnCO_3$ (c)	−195.20	Wagman et al. (1982)

Abbreviations see Table 1

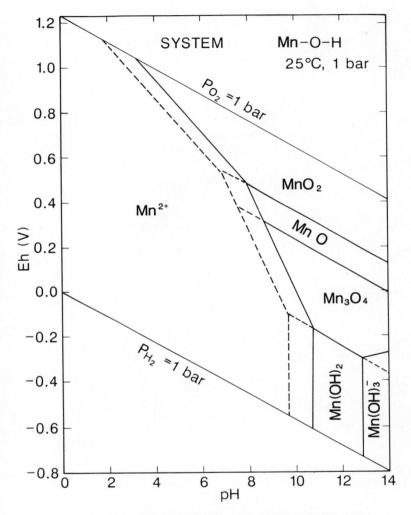

Fig. 49. Eh-pH diagram for part of the system Mn-O-H. Assumed activity for Mn = 10^{-6}. See text for discussion

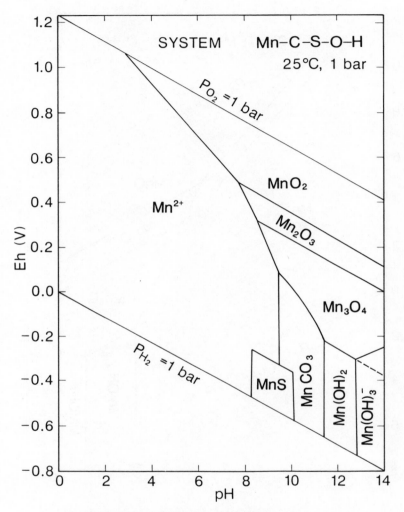

Fig. 50. Eh-pH diagram for part of the system Mn-C-S-O-H. Assumed activities for dissolved species are: Mn = 10^{-6}, C = 10^{-3}, S = 10^{-3}. See text for discussion

Technetium

The Eh-pH diagrams for technetium species are shown in Figs. 51 and 52. The thermodynamic data for important technetium species are given in Table 38.

Figure 51 shows the phase relations in the system Tc-O-H. Technetium possesses several stable valences in nature (O, II, III, IV, VI?, and VII). Of these, Tc(VII) in TcO_4^- is of extreme interest because of long-lived ^{99}Tc produced during nuclear fission reactions (half-life, 2.13×10^5 years). If TcO_4^- is formed, it is not sorbed readily on most geologic media (Brookins 1984). However, as shown in Fig. 51, if conditions are even mildly reducing, but still at Eh well above the sulfide-sulfate boundary, then Tc may be immobile as one or more Tc-oxides may form. Not shown in Fig. 51 is a possible TcO_3 stability field because of uncertain data. If TcO_3 falls between the TcO_2 to Tc_3O_4 and TcO_4^- fields, then the field of TcO_4^- is even further decreased.

In Fig. 52 we see the effect of adding S to this system. Here, a field of TcS_2 replaces the field of $Tc(OH)_2$ and native Tc (Fig. 51). The stability of the TcS_2 species is well documented under sulfur-present conditions in laboratory experiments, and may further retard and retain any escaping Tc from breached radioactive waste canisters.

Technetium contains only radioactive species in nature, hence use of chemically similar elements such as Re must be used to comment on Tc migration (see Rhenium). Previous Eh-pH diagrams for Tc species include those by Brookins (1978a, c; 1986b, 1987a), Khalil and White (1984), and Bird and Lopata (1980).

Table 38. Thermodynamic data for technetium

Species (state)	ΔG_f^0 (kcal/gfw)	Reference
TcOH (c)	−56.09	Rard (1983)
$Tc(OH)_2$ (c)	−110.20	Rard (1983)
Tc_3O_4 (c)	−206.40	Rard (1983)
TcO_4^{1-} (aq)	−149.10	Rard (1983)
TcS_2 (c)	−51.03	Rard (1983)
TcO_2 (c)	−88.29	Pourbaix (1966)

Abbreviations see Table 1

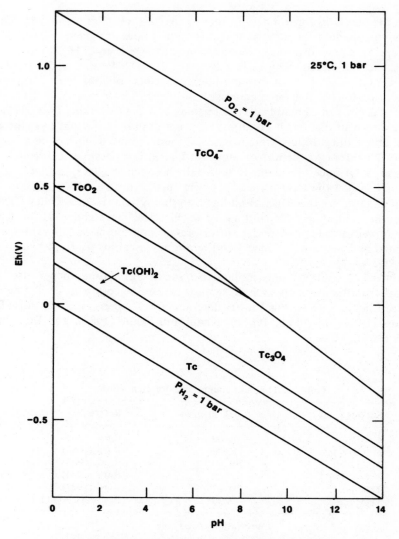

Fig. 51. Eh-pH diagram for part of the system Tc-O-H. Assumed activity for dissolved Tc = 10^{-8}. See text for discussion

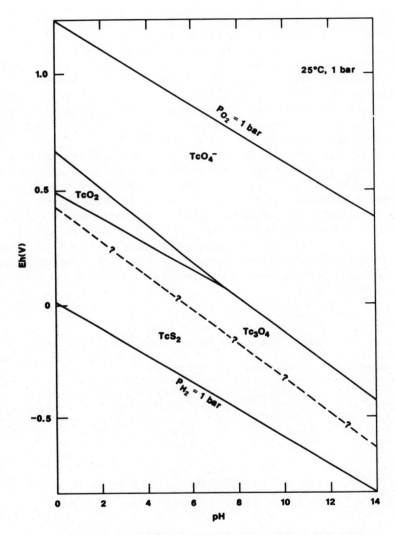

Fig. 52. Eh-pH diagram for part of the system Tc-S-O-H. Assumed activities for dissolved species are: Tc = 10^{-8}, S = 10^{-3}. See text for discussion

RHENIUM

The Eh-pH diagrams for rhenium species are shown in Figs. 53 and 54. The thermodynamic data for important rhenium species are given in Table 39.

Figure 53 shows the stability fields for phases in the system Re-O-H. Similar to Tc, the important species of Re(VII) is soluble ReO_4^- ion, which occupies much of the high-Eh part of the diagram at all pH values. Important Re valences are (0, III, IV, VI, and VII). Under mildly reducing conditions, and above the sulfide-sulfate boundary, Re(VII) is reduced first to Re(VI) and then to Re(IV,III) oxides. Native Re forms at lower Eh.

With the addition of sulfur, the phases in the system Re-S-O-H are shown in Fig. 54. Here, the field of native Re is replaced by ReS_2 and most of the Re_2O_3 and some of the ReO_2 and ReO_3 fields are replaced by ReS_2 as well. Not shown are phases such as Re_3O_4 and Re_2S_3 for which only questionable data exist. The total space occupied by solids as opposed to aqueous species in both the Tc($\pm$S)-O-H and Re-($\pm$S)-O-H systems is essentially the same. Further, Re and Tc possess the same valences and exhibit almost identical crystal chemical behavior (Brookins 1984), hence Re is often used as an analog for Tc in predicting Tc behavior from hypothetically breached radioactive wastes (Brookins 1986b, 1987a).

Table 39. Thermodynamic data for rhenium

Species (state)	ΔG_f^0 (kcal/gfw)	Reference
Re_2O_3 (c)	−138.64	Zoubov and Pourbaix (1966)
ReO_2 (c)	−87.95	Wagman et al. (1982)
ReO_3 (c)	−127.96	Calculated[a]
ReO_4^{1-} (aq)	−165.99	Wagman et al. (1982)
ReS_2 (c)	−38.10	Calculated[a]

Abbreviations see Table 1
[a] Method of calculation described in Brookins (1983a)

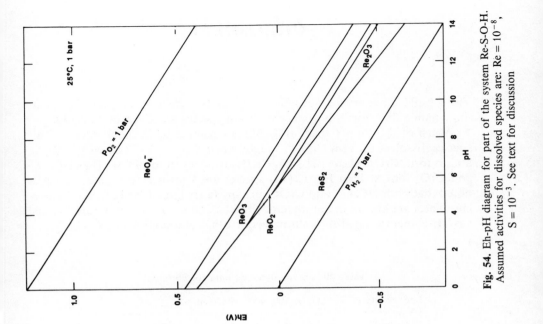

Fig. 54. Eh-pH diagram for part of the system Re-S-O-H. Assumed activities for dissolved species are: Re = 10^{-8}, S = 10^{-3}. See text for discussion

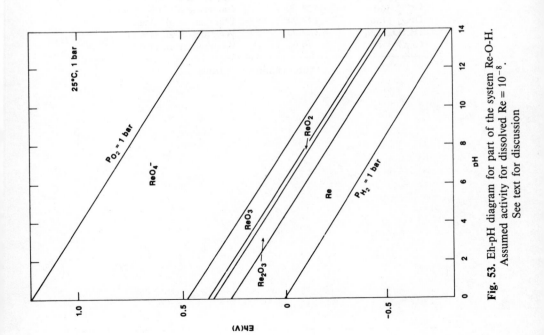

Fig. 53. Eh-pH diagram for part of the system Re-O-H. Assumed activity for dissolved Re = 10^{-8}. See text for discussion

CHROMIUM

The Eh-pH diagram for chromium species is shown in Fig. 55. The thermodynamic data for important chromium species are given in Table 40.

Much of the Eh-pH space (Fig. 55) is occupied by insoluble Cr_2O_3. This species dissolves to form $CrOH^{2+}$ slightly below pH = 5 (Cr activity = 10^{-6}), and to form CrO_2^- above pH 13.5. Cr(III) oxidizes to form Cr(VI) as $HCrO_4^-$ and CrO_4^{2-} ions at high Eh. Cr(VI) species are known carcinogens, and it is noted that both $HCrO_4^-$ and CrO_4^{2-} occupy fairly large Eh-pH fields. Various chromates are known in nature (crocoite, etc.). Cr_2O_3 is rare in nature as most Cr(III) is incorporated into chromites or other chromian spinels.

Table 40. Thermodynamic data for chromium

Species (state)	ΔG_f^0 (kcal/gfw)	Reference
Cr^{3+} (aq)	−51.50	Barner and Scheuerman (1978)
Cr_2O_3 (c)	−252.89	Wagman et al. (1982)
$Cr(OH)^{2+}$ (aq)	−103.00	Latimer (1952)
$Cr(OH)_2^+$ (aq)	−151.20	Garrels and Christ (1965)
CrO_2^- (aq)	−128.00	Garrels and Christ (1965)
CrO_4^{2-} (aq)	−173.94	Wagman et al. (1982)
$HCrO_4^-$ (aq)	−182.77	Wagman et al. (1982)
$Cr_2O_7^{2-}$ (aq)	−310.97	Wagman et al. (1982)

Abbreviations see Table 1

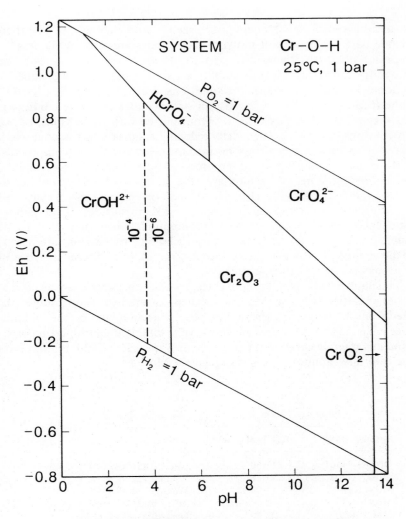

Fig. 55. Eh-pH diagram for part of the system Cr-O-H. Assumed activity of dissolved Cr = 10^{-6}. See text for discussion

MOLYBDENUM

The Eh-pH diagram for molybdenum species is shown in Fig. 56. The thermodynamic data for important molybdenum species are given in Table 41.

Important phases in the system Mo-S-O-H are shown in Fig. 56. The species MoS_2 occupies the sulfide-stable part of the diagram, and aqueous species dominate under sulfate-stable conditions, except for a small field of ilsemannite (Mo_3O_8). Mo species in natural waters under oxidizing conditions occur as various Mo(V, VI) oxyions. It is transported with similar ions, i.e., U, Se, V, As oxyions, and, when chemically reducing conditions are encountered, these various oxyions are replaced by insoluble, lower valence compounds. Thus, Mo is incorporated into the cryptocrystalline variety of molybdenite called jordisite (Brookins 1977). Similarly, the other ions yield separate compounds of other elements, i.e., U in pitchblende or coffinite, Se in native Se or in $Fe(S,Se)_2$, V in oxides or clay minerals, and As in sulfides. If subsequent oxidation encroaches on this group of compounds, then they tend to be segregated. Mo may be enriched by such a secondary process or, alternately, may be distributed but still enriched over background, thus making it an useful metal for geochemical U prospecting.

Molybdenum is a known toxin, and responsible for fairly widespread disease among cattle. Because Mo is enriched in uranium ores, some Mo separated from the ore during the milling process accumulates on the mill tailings piles area. As shown in Fig. 56, Mo is very soluble under high Eh conditions, and can migrate easily in the presence of water. Hence, mill tailings piles must be carefully monitored to ensure against Mo loss.

Table 41. Thermodynamic data for molybdenum

Species (state)	ΔG_f^0 (kcal/gfw)	Reference
$HMoO_4^-$ (aq)	−213.60	Garrels and Christ (1965)
MoO_4^{2-} (aq)	−199.88	Wagman et al. (1982)
MoS_2 (c)	−53.99	Wagman et al. (1982)
Mo_3O_8 (c)	−480.00	Titley and Anthony (1961)
MoO_2^+ (aq)	−122.80	Gerasimov et al. (1963)
MoO_2 (c)	−127.39	Wagman et al. (1982)
MoO_3 (c)	−159.65	Wagman et al. (1982)

Abbreviations see Table 1

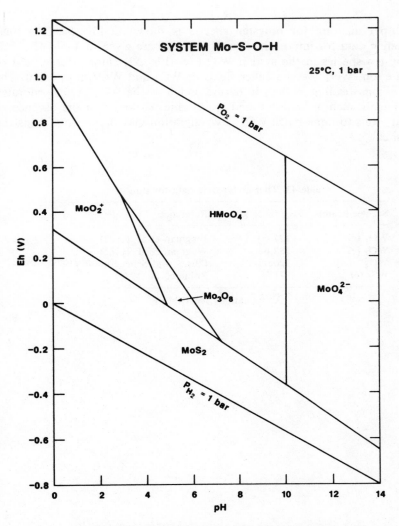

Fig. 56. Eh-pH diagram for part of the system Mo-S-O-H. Assumed activities for dissolved species are: Mo = 10^{-8}, S = 10^{-3}. See text for discussion

TUNGSTEN

The Eh-pH diagram for tungsten species is shown in Fig. 57. The thermodynamic data for important tungsten species are given in Table 42.

Tungsten species in the system W-O-H-S (Fig. 57) show a large field of tungstate ion, WO_4^{2-}, and smaller fields of WO_3 and Ws_2 (tungstenite). The diagram is misleading in that, in nature, much of the WO_4^{2-} is incorporated into minerals such as scheelite and wolframite. However, in the absence of necessary ions to immobilize WO_4^{2-}, W migration may occur from industrial waste sites.

Table 42. Thermodynamic data for tungsten

Species (state)	ΔG_f^0 (kcal/gfw)	Reference
WO_2 (c)	−127.60	Wagman et al. (1982)
WO_3 (c)	−182.60	Wagman et al. (1982)
WO_4^{2-} (aq)	−220.00	Barner and Scheuermann (1978)
WS_2 (c)	−71.20	Robie et al. (1978)

Abbreviations see Table 1

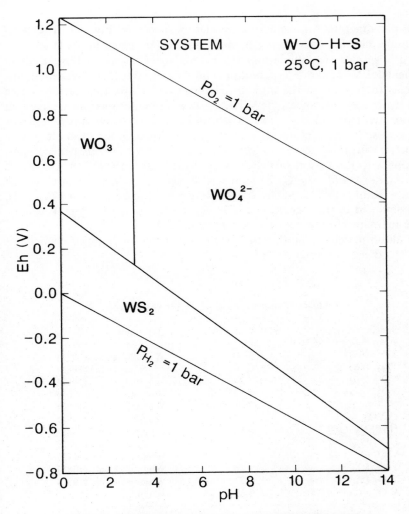

Fig. 57. Eh-pH diagram for part of the system W-O-H-S.
Assumed activities for dissolved species are: $W = 10^{-8}$, $S = 10^{-3}$.
See text for discussion

VANADIUM

The Eh-pH diagram for vanadium species is shown in Fig. 58. The thermodynamic data for important vanadium species are given in Table 43.

Most V species are soluble over the Eh-pH space shown in Fig. 58. Only small fields of V_2O_4 and V_2O_3 contain solid species. Yet, these oxides are important as they occur in the middle pH range and above the sulfide-sulfate boundary. In many instances, the V(III, IV) does not enter separate oxides (i.e., haggite or montroseite) but is rather incorporated into the octahedral sites of clay minerals. In the Grants Mineral Belt, New Mexico, for example, the V transported with the U enters chlorites and other clay minerals penecontemporaneous with the U minerals (Brookins 1976b, 1977, 1979a, 1984). This is also of interest should the Eh be increased in such a setting. While U will be oxidized to soluble U(VI) species, the V, while it may be oxidized to V(V < VI), is still contained in the clay minerals. Thus, segregation of U from V occurs, and the remnant V may be used as a pathfinder for U.

Vanadium toxicity is mild, and V is not identified as a major element of concern to the environment or to public health.

Table 43. Thermodynamic data for vanadium

Species (state)	ΔG_f^0 (kcal/gfw)	Reference
VO^{2+} (aq)	−106.69	Wagman et al. (1982)
VO_2^+ (aq)	−140.30	Wagman et al. (1982)
VO_3^- (aq)	−137.28	Wagman et al. (1982)
VO_4^{3-} (aq)	−214.87	Wagman et al. (1982)
HVO_4^{2-} (aq)	−232.98	Wagman et al. (1982)
$H_2VO_4^-$ (aq)	−243.98	Wagman et al. (1982)
$HV_{10}O_{28}^{5-}$ (aq)	−1840.80	Wagman et al. (1982)
$H_2V_{10}O_{28}^{4-}$ (aq)	−1845.84	Wagman et al. (1982)
V_2O_3 (c)	−272.30	Wagman et al. (1982)
V_2O_4 (c)	−315.08	Wagman et al. (1982)
V_2O_5 (c)	−339.27	Wagman et al. (1982)
$V_2O_7^{4-}$ (aq)	−410.85	Wagman et al. (1982)

Abbreviations see Table 1

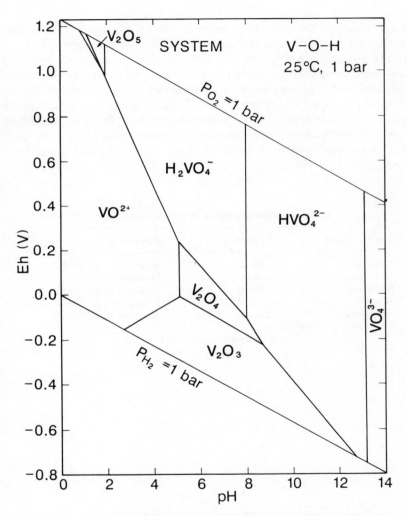

Fig. 58. Eh-pH diagram for part of the system V-O-H. Assumed activity for dissolved V = 10^{-6}. See text for discussion

NIOBIUM

The Eh-pH diagram for niobium species is shown in Fig. 59. The thermodynamic data for important niobium species are given in Table 44.

The only important Nb species is Nb(V) as Nb_2O_5. In nature, Nb(V) is commonly incorporated into niobate-tantalate salts. All are insoluble and are commonly enriched in placers upon rock weathering. A small field of Nb(III) as Nb^{3+} for an assumed Nb activity of 10^{-8} occurs under the most extreme acidic and lowest Eh is noted. The large stability field of Nb_2O_5 indicates lack of migration of Nb in nature, and it is noteworthy that Nb at the natural Oklo Reactor, Gabon, produced by fission, has been 100% retained in the host pitchblende (Brookins 1984).

Table 44. Thermodynamic data for niobium

Species (state)	ΔG_f^0 (kcal/gfw)	Reference
NbO (c)	−90.49	Wagman et al. (1982)
Nb^{3+} (aq)	−76.00	Latimer (1952)
NbO_2 (c)	−190.30	Wagman et al. (1982)
Nb_2O_5 (c)	−453.99	Wagman et al. (1982)
NbO_3^- (aq)	−222.78	Wagman et al. (1982)

Abbreviations see Table 1

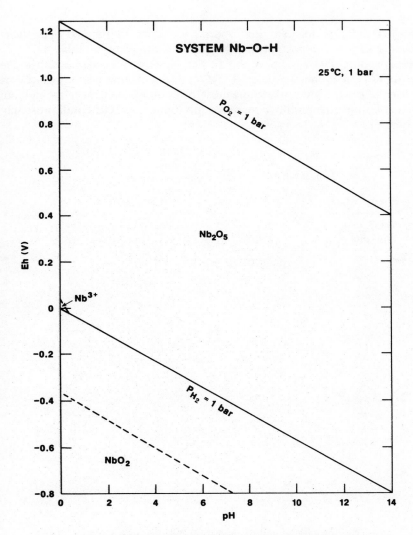

Fig. 59. Eh-pH diagram for part of the system Nb-O-H. Assumed activity of dissolved Nb = 10^{-8}. See text for discussion

TANTALUM

The Eh-pH diagram for tantalum species is shown in Fig. 60. The thermodynamic data for important tantalum species are given in Table 45.

Like Nb, only Ta(V) is important in nature. However, unlike Nb, in the system Ta-O-H a moderate field of TaO_2^+ occurs below pH = 5. Ta_2O_5 is stable over all Eh at higher pH. While most Ta is contained in tantalate-niobate salts, weathering under mildly to more pronounced acidic conditions may release Ta(V) as TaO_2^+.

Table 45. Thermodynamic data for tantalum

Species (state)	ΔG_f^0 (kcal/gfw)	Reference
TaO_2^+ (aq)	−201.39	Wagman et al. (1982)
Ta_2O_5 (c)	−456.79	Wagman et al. (1982)

Abbreviations see Table 1

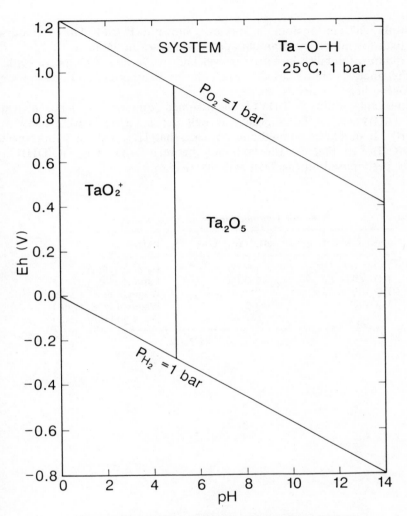

Fig. 60. Eh-pH diagram for part of the system Ta-O-H. Assumed activity of dissolved Ta = 10^{-8}. See text for discussion

TITANIUM

The Eh-pH diagram for titanium species is shown in Fig. 61. The thermodynamic data for important titanium species are given in Table 46.

Titanium insolubility in nature is well known. Various TiO_2 polymorphs (brookite, anatase) are remarkably stable during weathering, and concentrate into placers, heavy minerals, etc.

Interestingly, although Ti(IV) is the assumed common Ti valence, a field of the Ti(III)-bearing Ti_3O_5 occurs as well *if* the Ti(IV) bearing phase is $TiO(OH)_2$. It should be pointed out, however, that TiO_2 *is* stable with respect to $TiO(OH)_2$. In Fig. 61, the boundary between Ti_3O_5 and $TiO(OH)_2$ is shown superimposed on the TiO_2 stability field.

Table 46. Thermodynamic data for titanium

Species (state)	ΔG_f^0 (kcal/gfw)	Reference
TiO^{2+} (aq)	−138.00	Latimer (1952)
$TiO \cdot (OH)_2$ (c)	−253.00	Latimer (1952)
TiO_2 (c, anatase)	−211.40	Wagman et al. (1982)
Ti_2O_3 (c)	−342.78	Wagman et al. (1982)
Ti_3O_5 (c)	−553.87	Wagman et al. (1982)

Abbreviations see Table 1

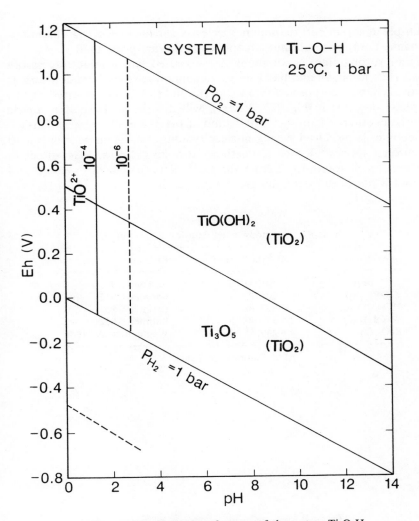

Fig. 61. Eh-pH diagram for part of the system Ti-O-H.
Assumed activity of dissolved Ti = 10^{-6}.
Note: TiO$_2$ occupies *all* Eh-pH space shown as TiO(OH)$_2$ and Ti$_3$O$_5$.
See text for discussion

ZIRCONIUM

The Eh-pH diagram for zirconium species is shown in Fig. 62. The thermodynamic data for important zirconium species are given in Table 47.

In nature, zirconium is commonly incorporated into insoluble accessory minerals such as zircon and baddeleyite. Zirconium released by rock weathering may likely be incorporated into $Zr(OH)_4$, and this is the preferred Zr(IV) solid phase depicted in Fig. 62. $Zr(OH)_4$ will dissolve to form Zr^{4+} under acidic pH less than 1.2 for $a_{Zr} = 10^{-6}$, and at pH above 13 to form $HZrO_3^-$.

Zirconium is produced during nuclear fission, and ^{99}Zr is a precursor to the important species ^{99}Tc (see Technetium). Brookins (1983a, 1984) discusses Zr systematics in some detail from the Oklo Natural Reactor, Gabon, and notes that it has been essentially 100% retained at the site.

Table 47. Thermodynamic data for zirconium

Species (state)	ΔG_f^0 (kcal/gfw)	Reference
Zr^{4+} (aq)	−141.00	Latimer (1952)
ZrO^{2+} (aq)	−200.90	Latimer (1952)
$HZrO_3^-$ (aq)	−287.70	Latimer (1952)
$Zr(OH)_4$ (c)	−370.00	Latimer (1952)
ZrO_2 (c)	−249.23	Wagman et al. (1982)

Abbreviations see Table 1

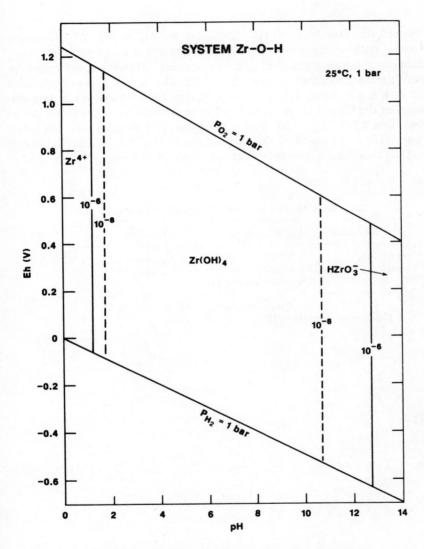

Fig. 62. Eh-pH diagram for part of the system Zr-O-H. Assumed activity of dissolved Zr = 10^{-6}. See text for discussion

HAFNIUM

The Eh-pH diagram for hafnium species is shown in Fig. 63. The thermodynamic data for important hafnium species are given in Table 48.

Hafnium in nature is camouflaged by zirconium. Thermodynamic data for Hf species are not as complete as those for Zr. This is reflected in both Table 48 and in Fig. 63. Using HfO_2 as the preferred form of Hf(IV), it is noted that most of the Eh-pH space is covered by this phase. Only at extemely acidic pH (less than 0.2 for $a_{Hf} = 10^{-8}$) does HfO_2 dissolve to form Hf^{4+}. No data exist for $HHfO_3^-$ but based on the diagram for Zr (Fig. 62), it is probable that some HfO_2 dissolution also occurs at very high pH as well.

Table 48. Thermodynamic data for hafnium

Species (state)	ΔG_f^0 (kcal/gfw)	Reference
HfO_2 (c)	−260.09	Wagman et al. (1982)
Hf^{4+} (aq)	−156.80	Latimer (1952)[a]
	No data for $Hf(OH)_5^-$	

Abbreviations see Table 1
[a] Calculated from $Hf \rightleftharpoons Hf^{4+} + 4e^-$, $E^0 = 1.70$

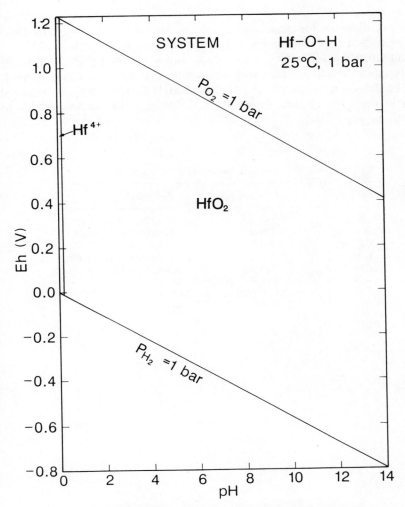

Fig. 63. Eh-pH diagram for part of the system Hf-O-H. Assumed activity of dissolved Hf = 10^{-8}. See text for discussion

SCANDIUM

The Eh-pH diagram for scandium species is shown in Fig. 64. The thermodynamic data for important scandium species are given in Table 49.

Scandium often follows Fe^{3+} in nature, and it is noted that the field of Sc_2O_3 occupies much of the space of hematite, Fe_2O_3 (see Figs. 36–42). Unlike Fe, however, Sc possesses only the Sc(III) valence, and it forms no sulfides in the stability field of water. Sc^{3+} forms from the sesquioxide at pH = 3.9 ($a_{Sc} = 10^{-6}$).

Table 49. Thermodynamic data for scandium

Species (state)	ΔG_f^0 (kcal/gfw)	Reference
Sc^{3+} (aq)	−140.20	Wagman et al. (1982)
$ScOH^{2+}$ (aq)	−191.49	Wagman et al. (1982)
$Sc(OH)_3$ (c)	−294.77	Wagman et al. (1982)
Sc_2O_3 (c)	−434.84	Wagman et al. (1982)
$Sc(OH)_4^-$ (aq)	−327.60	Baes and Mesmer (1976)

Abbreviations see Table 1

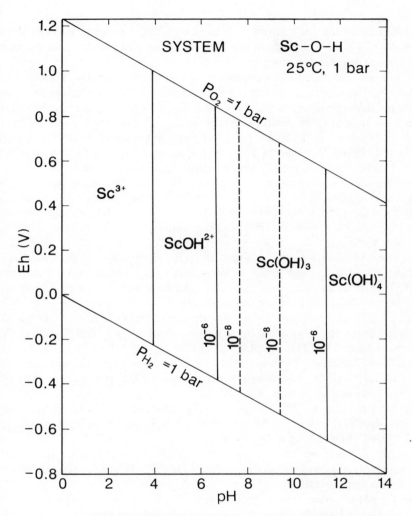

Fig. 64. Eh-pH diagram for part of the system Sc-O-H. Assumed activity of dissolved Sc = 10^{-8}. See text for discussion

YTTRIUM AND THE RARE EARTH ELEMENTS (REE)

The Eh-pH diagrams, for yttrium and the rare earth element species are shown in Figs. 65–79. The thermodynamic data for important yttrium and the rare earth element species are given in Table 50.

Yttrium and the rare earth elements [Lu, Yb, Tm, Er, Ho, Dy, Tb, Gd, Eu, Sm, Nd, Pr, Ce, La (note: Pm contains only short-lived radioactive isotopes and no reliable thermodynamic data exist for its species)] are treated here together because of their similar crystal-chemical and solution-chemical properties. The data of Table 50 are compiled from Schumm et al. (1973), Smith and Martell (1976), Wagman et al. (1982), Baes and Mesmer (1976), and the author (Brookins 1983). Each element will be discussed in turn.

Yttrium species are shown in Fig. 65. The important solid species are $Y(OH)_3$ and $Y_2(CO_3)_3$. Y^{3+} is the dominant aqueous species. The aqueous species $Y(OH)_4^-$ plots above pH = 14 and is not shown. The carbonate phase is important as otherwise $Y(OH)_3$ would dissolve to Y^{3+} at higher pH than shown for $Y_2(CO_3)_3$:Y^{3+}.

The Eh-pH diagram for Lu species (Fig. 66) shows large fields for $Lu(OH)_3$ and Lu^{3+}, and only a very small, but important, field for $Lu_2(CO_3)_3$. The aqueous species $Lu(OH)_4^-$ probably plots in the Eh-pH space but no thermodynamic data are available for it.

The Eh-pH diagram for Yb species (Fig. 67) shows dominant fields of Yb^{3+} and $Yb(OH)_3$, separated by a narrow field of $Yb_2(CO_3)_3$. $Yb(OH)_3$ dissolves to form $Yb(OH)_4^-$ at pH = 12 ($a_{Yb} = 10^{-6}$).

The Eh-pH diagram for Tm species (Fig. 68) also shows large fields for Tm^{3+} and $Tm(OH)_3$ separated by a narrow field of $Tm_2(CO_3)_3$. No field of $Tm(OH)_4^-$ is plotted due to lack of data.

The Eh-pH diagram for Er species (Fig. 69) also shows large fields of Er^{3+} and $Er(OH)_3$ separated by a field of $Er_2(CO_3)_3$. $Er(OH)_3$ dissolves to form $Er(OH)_4^-$ at pH = 11.6 ($a_{Er} = 10^{-6}$).

The Eh-pH diagram for Ho species (Fig. 70) contains large fields for Ho^{3+} and $Ho(OH)_3$ with a narrow field of $Ho_2(CO_3)_3$ in between. Data are not available for $Ho(OH)_4^-$.

The Eh-pH diagram for Dy species (Fig. 71) again shows large fields for Dy^{3+} and $Dy(OH)_3$ with the usual intermediate field of $Dy_2(CO_3)_3$. At high pH (near 11.7, $a_{Dy} = 10^{-6}$) $Dy(OH)_3$ dissolves to form $Dy(OH)_4^-$.

The Eh-pH diagram for Tb species (Fig. 72) differs from the diagrams for Lu to Dy (Figs. 66–71) in that the fields of Tb^{3+} and $Tb(OH)_3$ are separated by a more pronounced field of $Tb_2(CO_3)_3$, which is characteristic of the middle rare earth elements (see Brookins 1983). No data are available for $Tb(OH)_4^-$.

Table 50. Thermodynamic data[a] [ΔG_f^0 (kcal/gfw)] for the rare earth elements and yttrium A

(a) M(III) species

Element	M^{3+} (aq)	$M(OH)_3$ (c)	M_2O_3 (c)	$M_2(CO_3)_3$ (c)	$M(OH)_4^-$ (aq)
La	−163.4	(−305.8)	−407.7	−750.9	−
Ce	−160.6	(−303.6)	−407.8	(−744.1)	−
Pr	−162.3	−307.1	_____	(−747.1)	−
Nd	−160.5	(−305.2)	−411.3	−741.4	−336.7
Sm	−159.3	(−306.9)	−414.6	−741.4	−
Eu	−137.2	−285.5	−372.2	(−697.3)	−
Gd	−158.0	(−306.9)	_____	(−738.9)	−338.0
Tb	−155.8	(−303.4)	_____	(−734.5)	−
Dy	−159.0	(−307.2)	−423.4	−739.3	−339.9
Ho	−161.0	(−310.1)	−428.1	(−744.9)	−
Er	−159.9	(−309.5)	−432.3	(−742.7)	−342.2
Tm	−158.2	(−307.8)	−428.9	(−739.3)	−
Yb	−153.9	(−303.9)	−412.6	−729.1	−336.0
Lu	−150.0	(−300.3)	−427.6	(−722.9)	−
Y	−165.8	−308.6	−434.2	−752.4	−341.2

(b) Other rare earth element data

Species (state)	ΔG_f^0 (kcal/gfw)	Reference
Eu^{2+} (aq)	−129.10	Wagman et al. (1982)
CeO_2 (c)	−244.40	Wagman et al. (1982)
Ce^{4+} (aq)	−120.44	Wagman et al. (1982)

[a] See Brookins (1983) for discussion of errors. Data in parentheses () calculated by Brookins (1983), the remainder of the data are from Schumm et al. (1973) and Smith and Martell (1976). Abbreviations see Table 1

The Eh-pH diagram for Gd species (Fig. 73) is similar to that for Tb (Fig. 72) in that Gd^{3+} is separated from $Gd(OH)_3$ by a large field for $Gd_2(CO_3)_3$. $Gd(OH)_3$ dissolves to form $Gd(OH)_4^-$ at pH near 12.8 ($a_{Gd} = 10^{-6}$).

The Eh-pH diagram for Eu species (Fig. 74) is markedly different from the other REE because of the presence of the Eu(II) species, Eu^{2+}. Figure 74 is constructed to show the field of Eu^{2+}. Hence, the field of $Eu_2(CO_3)_3$ which separates Eu^{3+} from $Eu(OH)_3$ appears very narrow, but note that this results from taking $a_{Eu} = 10^{-8}$. The dashed line for $a_{Eu} = 10^{-6}$ should be used for comparison to Figs. 72 and 73, for example. Note that if only $a_{Eu} = 10^{-6}$ is used, the field of Eu^{2+} has virtually disappeared. Data are not available for $Eu(OH)_4^-$ nor for $EuCO_3$.

The Eh-pH diagram for Sm species (Fig. 75) is similar to those for Tb and Gd (Figs. 72, 73) with a substantial field of $Sm_2(CO_3)_3$ separating Sm^{3+} from $Sm(OH)_3$. Again, no data for the $Sm(OH)_4^-$ species are available.

The Eh-pH diagram for Nd species (Fig. 76) also shows a large field for $Nd_2(CO_3)_3$ in between the fields for Nd^{3+} and $Nd(OH)_3$. $Nd(OH)_3$ dissolves to form $Nd(OH)_4^-$ at pH near 12.5 ($a_{Nd} = 10^{-6}$).

The Eh-pH diagram for Pr species (Fig. 77) shows large fields for Pr^{3+} and $Pr_2(CO_3)_3$ as well as $Pr(OH)_3$. No data are available for $Pr(OH)_4^-$.

The Eh-pH diagram for Ce species (Fig. 78) is different from the other REE Eh-pH diagrams. Ce possesses a Ce(IV) valence in addition to Ce(III), and large fields for the Ce(IV) species CeO_2 dominates much of the water stability field. This field also truncates much of the Ce^{3+} field as well. The importance of the CeO_2 field is discussed later in this section. No data are available for $Ce(OH)_4^-$. The fields of both $Ce_2(CO_3)_3$ and $Ce(OH)_3$ are displaced by the large field of CeO_2.

The Eh-pH diagram for La species (Fig. 79) shows large stability fields for La^{3+}, $La_2(CO_3)_3$, and $La(OH)_3$. No data are available for $La(OH)_4^-$.

Discussion of the Yttrium and REE Eh-pH Diagrams

The REE, in general, fall into three groups based on the Eh-pH diagrams: La-Gd, Tb-Ho, and Yb-Lu, based on the relative importance of $M_2(CO_3)_3$ stability. Of the first group, only Nd shoes a reduced field of $Nd_2(CO_3)_3$ relative to the other La-Gd carbonates. This reduced $Nd_2(CO_3)_3$ field may account, for example, for the limited Nd migration noted at the Oklo Natural Reactor (Curtis et al. 1983), while the other REE were immobile. Without the presence of the $M_2(CO_3)_3$ phases (i.e., in a carbonate-free to -poor environment), the boundaries between $M^{3+}:M(OH)_3$ fall on the alkaline side of neutrality (pH values 7–8 except for Yb and Lu). Thus, in sea water the REE might be more mobile than actually noted were it not for this carbonate phase (Brookins 1983b).

Europium and Ce are the only two REE with valences other than (III). The importance of Eu(II) in petrology is well known (Henderson 1984), especially in moderate to high temperature settings. Under extremely reducing conditions, Eu(II) may be important in keeping Eu mobile, while the other REE are insoluble. This remains to be tested in detail.

The REE are depleted in sea water (Piper 1974) owing to the extreme insolubility of M:(OH)$_3$ compounds and, for Ce, insolubility of CeO$_2$. Yet, Ce is much more depleted than the other REE in most sea water which, if considered just in light of solubility product constants, should not necessarily be the case.

Elderfield and Greaves (1982) have also shown that Ce is highly depleted in most ocean waters relative to the other REE, which they attribute to the probable oxidation of Ce(III) to Ce(IV) with removal as CeO$_2$. The diagram presented here (Fig. 78) clearly supports this interpretation, although the diagrams for the REE (Figs. 65–79) fail to provide any information for the relative enrichment of surface waters in the REE relative to deeper waters (see discussion in Elderfield and Greaves 1982). Further, their work also shows a lack of correlation with the REE with either P or Si. This is important since P-ligands as transporting agents for the REE (as $M_2P_2O_7^{2+}$?) which have been discussed by the author (Brookins 1979a) may compete with the simple M^{3+} ions for typical ocean water P concentrations. Yet, total P increases from surface waters with depth while total REE decrease, which suggests that P-ligands are not the prime species responsible for the REE behavior (Elderfield and Greaves 1982). Surface water enrichment in colloids, organic life, amorphous coatings, or other factors may be responsible for the difference between surface water and deeper water (Elderfield and Greaves 1982).

Inspection of the Eh-pH diagram for Ce relative to the other REE shows that the boundary between Ce^{3+} and CeO_2 effectively reduces the Eh-pH limits of Ce^{3+} relative to other REE-M^{3+} species. Very simply, this means that any Ce^{3+} available from weathering or any other source will be oxidized to Ce(IV) and removed as CeO$_2$, over virtually any slight pH shift, whereas M(III) REE removal is restricted to pH values of about 7.5 and higher. The field for $M_2(CO_3)_3$ is important for the REE as it argues for REE removal at pH values less than that indicated by the simple M^{3+} : M(OH)$_3$ boundary. The REE are commonly described as precipitating in oxide-hydroxide phases on the seafloor, but the data and diagrams presented here suggest the possibility of incorporation into carbonate phases as well.

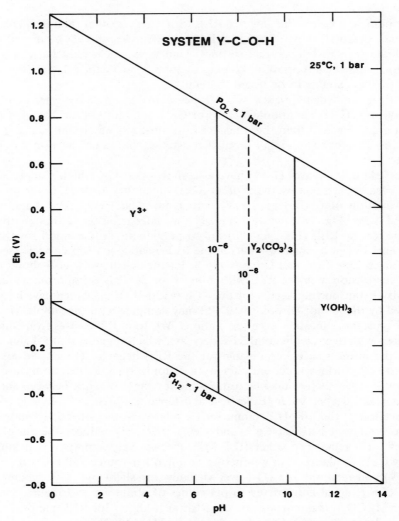

Fig. 65. Eh-pH diagram for part of the system Y-C-O-H. Assumed activities of dissolved species are: $Y = 10^{-6, -8}$, $C = 10^{-3}$. See text for discussion

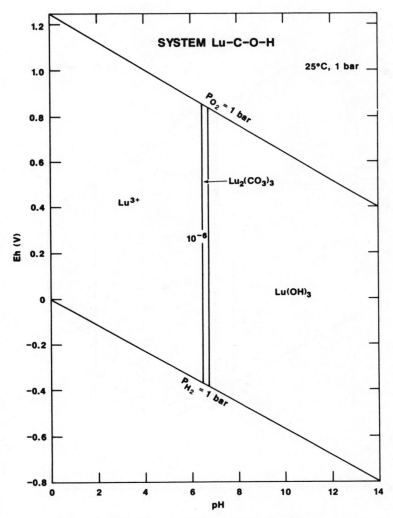

Fig. 66. Eh-pH diagram for part of the system Lu-C-O-H. Assumed activities for dissolved species are: Lu = $10^{-8, -6}$, C = 10^{-3}. See text for discussion

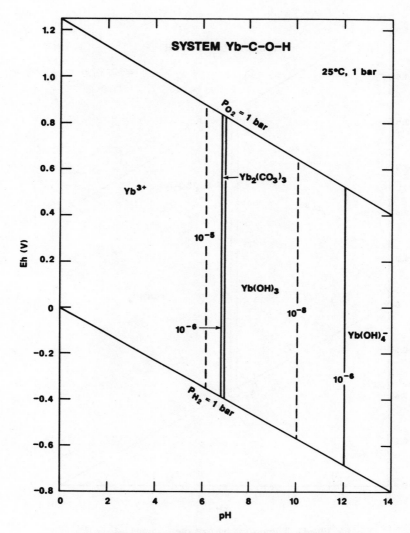

Fig. 67. Eh-pH diagram for part of the system Yb-C-O-H. Assumed activities for dissolved species are: $Yb = 10^{-6,\ -8}$, $C = 10^{-3}$. See text for discussion

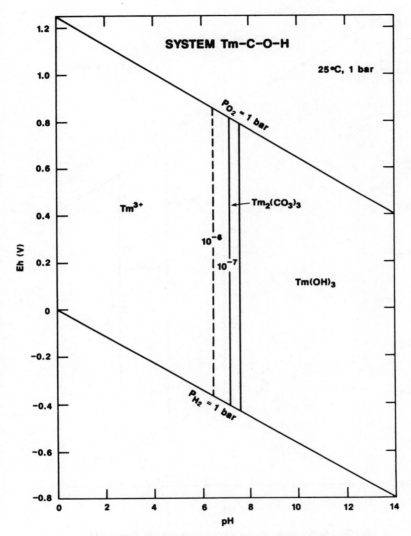

Fig. 68. Eh-pH diagram for part of the system Tm-C-O-H. Assumed activities for dissolved species are: Tm = $10^{-7, -6}$, C = 10^{-3}. See text for discussion

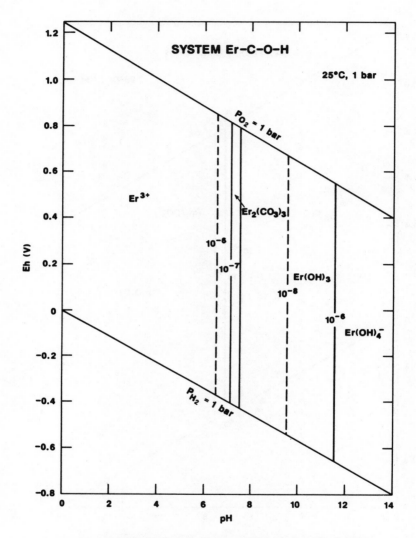

Fig. 69. Eh-pH diagram for part of the system Er-C-O-H. Assumed activities for dissolved species are: Er = $10^{-8, -6}$, C = 10^{-3}. See text for discussion

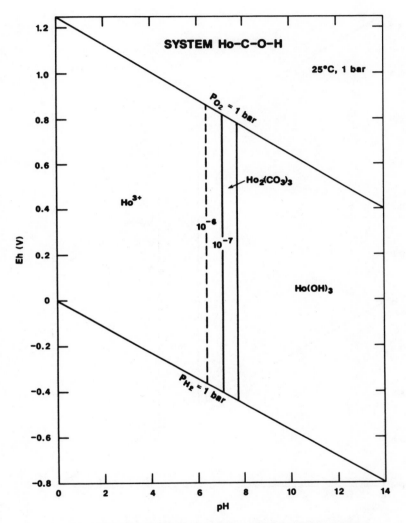

Fig. 70. Eh-pH diagram for part of the system Ho-C-O-H.
Assumed activities for dissolved species are: Ho = $10^{-7,\ -6}$, C = 10^{-3}.
See text for discussion

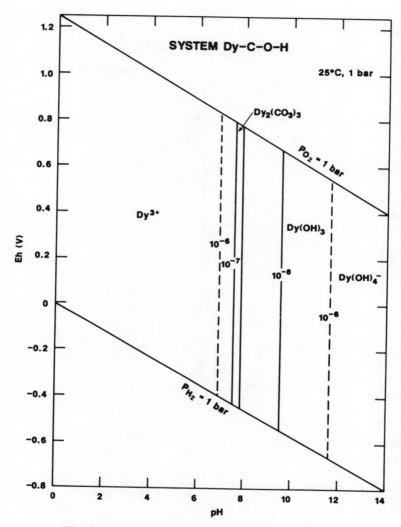

Fig. 71. Eh-pH diagram for part of the system Dy-C-O-H. Assumed activities for dissolved species are: Dy = $10^{-6, -7}$, C = 10^{-3}. See text for discussion

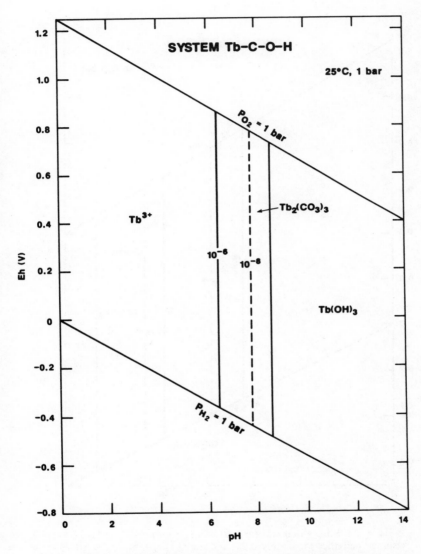

Fig. 72. Eh-pH diagram for part of the system Tb-C-O-H. Assumed activities for dissolved species are: Tb = $10^{-6, -8}$, C = 10^{-3}. See text for discussion

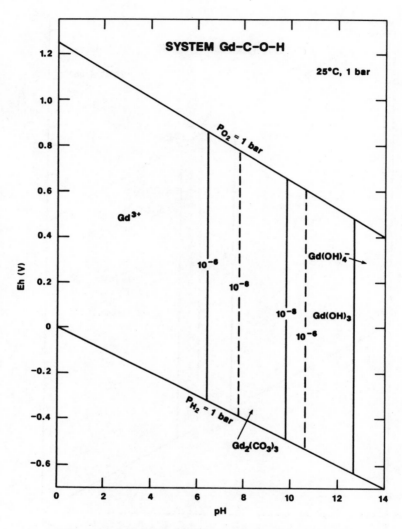

Fig. 73. Eh-pH diagram for part of the system Gd-C-O-H. Assumed activities for dissolved species are: Gd = $10^{-8, -6}$, C = 10^{-3}. See text for discussion

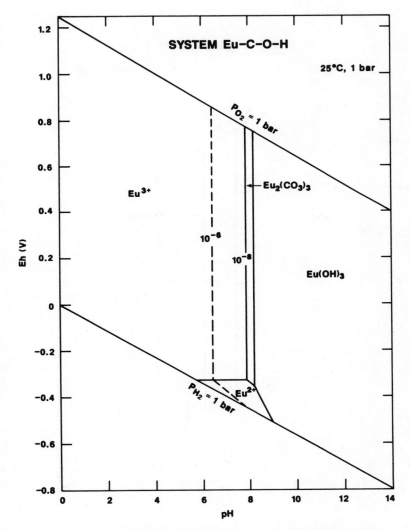

Fig. 74. Eh-pH diagram for part of the system Eu-C-O-H. Assumed activities for dissolved species are: Eu = $10^{-8, -6}$, C = 10^{-3}. See text for discussion

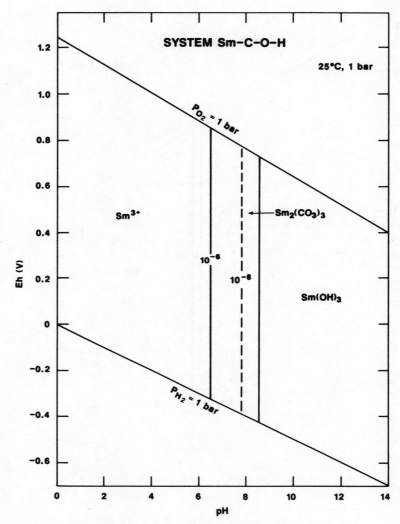

Fig. 75. Eh-pH diagram for part of the system Sm-C-O-H. Assumed activities for dissolved species are: Sm = $10^{-6, -8}$, C = 10^{-3}. See text for discussion

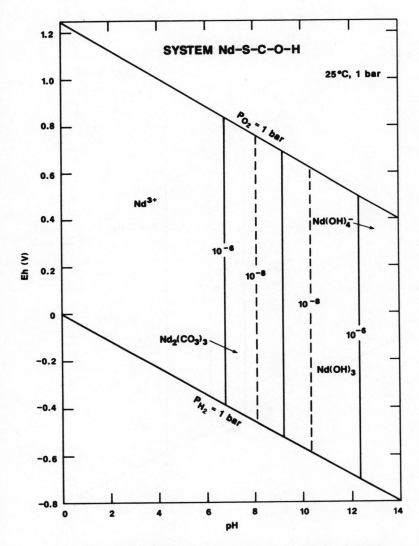

Fig. 76. Eh-pH diagram for part of the system Nd-S-C-O-H. Assumed activities for dissolved species are: Nd = $10^{-6,\ -8}$, C = 10^{-3}. See text for discussion

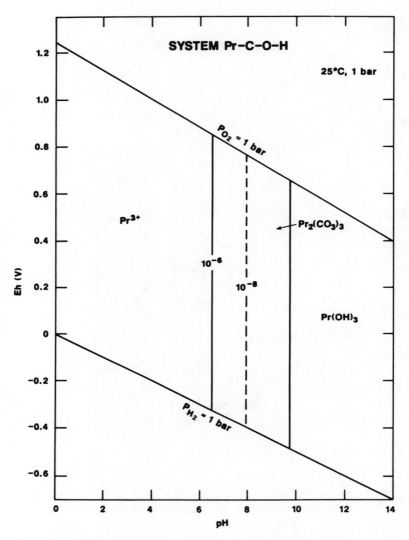

Fig. 77. Eh-pH diagram for part of the system Pr-C-O-H.
Assumed activities for dissolved species are: Pr = $10^{-8, \ -6}$, C = 10^{-3}.
See text for details

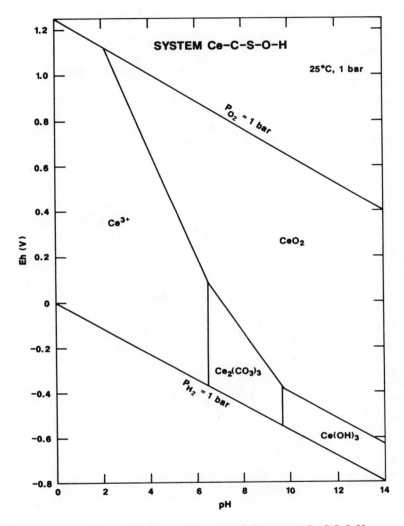

Fig. 78. Eh-pH diagram for part of the system Ce-C-S-O-H. Assumed activities for dissolved species are: $Ce = 10^{-8, -6}$, $C = 10^{-3}$
See text for details

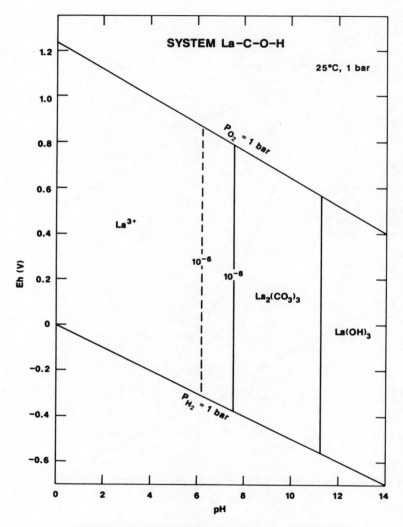

Fig. 79. Eh-pH diagram for part of the system La-C-O-H. Assumed activities for dissolved species are: La = $10^{-6, -8}$, C = 10^{-3}. See text for discussion

AMERICIUM

Note: For Am, Pu, Np and U there is an unresolved question as to whether or not $M(OH)_5^-$ species actually exist. The OECD (1985) data list values for the $M(OH)_5^-$ species for Am, Pu, Np and U, however, and the stability fields for these species are shown accordingly in the following diagrams.

The Eh-pH diagrams for americium species are shown in Figs. 80 and 81. The thermodynamic data for important americium species are given in Table 51.

The Eh-pH phase relations in the system Am-O-H are shown in Fig. 80. Large amounts of Eh-pH space are covered by aqueous Am^{3+} and $Am(OH)_5^-$, while the solids, $Am(OH)_3$ and AmO_2, occupy smaller fields. Interestingly, under oxidizing, near-neutral pH, Am is immobilized as AmO_2, yet under more reducing conditions, again at pH near 7, Am^{3+} may form and thus Am may be highly mobile. This is of concern to Am retention in buried radioactive wastes.

In Fig. 81, the importance of dissolved CO_2 is shown in the system Am-C-O-H. Here, it is noted that a large stability field of $Am_2(CO_3)_3$ appears at the expense of Am^{3+}, AmO_2, and $Am(OH)_3$. With reference to buried radioactive wastes, this suggests that the presence of carbonate in the repository host rocks will be beneficial as it provides an immobilization barrier to Am. More complex Am carbonate species are noted (OECD 1985), but their role is uncertain (see discussion in Brookins 1984).

Previously published Eh-pH diagrams for Am species include those by Brookins (1978c, 1979b, 1984). Those diagrams were prepared using older data, and the data of Table 51 are the currently preferred values.

Table 51. Thermodynamic data for americium

Species (state)	ΔG_f^0 (kcal/gfw)	Reference
Am^{3+} (aq)	−143.19	OECD (1985)
$Am_2(CO_3)_3$ (c)	−716.11	OECD (1985)
$Am(OH)_3$ (c)	−279.16	OECD (1985)
AmO_2 (c)	−210.43	OECD (1985)
$Am(OH)_3$ (am)	−289.38	OECD (1985)
Am^{4+} (aq)	−89.20	OECD (1985)
Am_2O_3 (c)	−385.97	OECD (1985)
$Am(OH)_5^-$ (aq)	−356.58	OECD (1985)
$AmOH^{2+}$ (aq)	−189.65	OECD (1985)

Abbreviations see Table 1

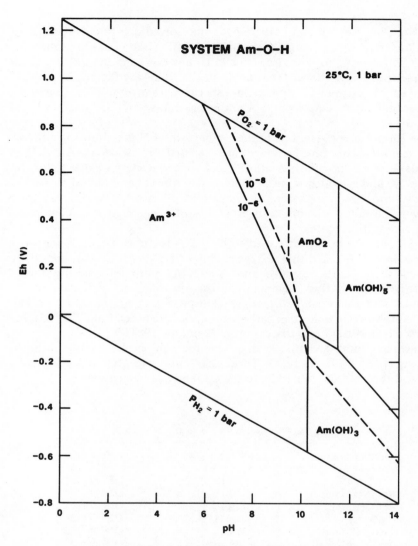

Fig. 80. Eh-pH diagram for part of the system AM-O-H. Assumed activity of dissolved Am = $10^{-6, -8}$. See text for discussion

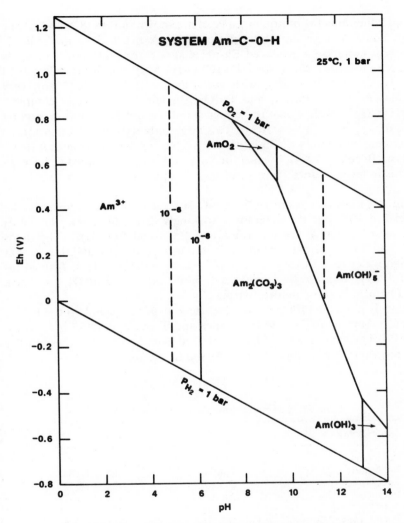

Fig. 81. Eh-pH diagram for part of the system Am-C-O-H. Assumed activities for dissolved species are: Am = $10^{-6, -8}$, C = 10^{-3}. See text for discussion

PLUTONIUM

The Eh-pH diagrams for plutonium species are shown in Figs. 82 and 83. The thermodynamic data for important plutonium species are given in Table 52.

The most commonly produced Eh-pH diagram for plutonium species is shown in Fig. 82 for the system Pu-(C)-O-H. Most of the stability field of water is covered by PuO_2, with subordinate fields of Pu^{3+} and very small fields of Pu^{4+}, PuO_2^+, and $Pu(OH)_5^-$. The Pu-carbonate complex, $PuO_2(OH)_2CO_3^{2-}$, plots above the upper stability limit for water (assuming activities of: $a_{Pu} = 10^{-8}$, $a_C = 10^{-3}$). This large stability field of PuO_2 attests to the extreme immobility of Pu in nature, such as at the Oklo Natural Reactor, Gabon (Brookins 1984). Further, activity of dissolved Pu must be lowered to unrealistic levels before there is an appreciable diminishing of the PuO_2 field.

In Fig. 83, however, in the system Pu-O-H-C, if $Pu(OH)_4$ is chosen as the preferred Pu(IV) solid, the diagram is strikingly different. Here, the solids $Pu(OH)_4$ and $Pu_2(CO_3)_3$ are subordinate to Pu^{3+} and $Pu(OH)_5^-$. The fields of Pu^{4+} and PuO_2^+ remain small. The importance of this diagram is that under high water to rock ratios, Pu may be mobile if $Pu(OH)_4$ is formed instead of PuO_2. It must be emphasized, however, that $Pu(OH)_4$ will age quickly to PuO_2 in most natural settings.

Previously published Eh-pH diagrams for Pu species include those by the author (Brookins 1978b, c, 1984), Lemire and Tremain (1980), and Krauskopf (1986). The data of Table 52 are preferred over data used in these earlier studies, however; hence, Figs. 82 and 83 should be referenced hereon.

Table 52. Thermodynamic data for plutonium

Species (state)	ΔG_f^0 (kcal/gfw)	Reference
Pu^{3+} (aq)	−138.15	OECD (1985)
Pu^{4+} (aq)	−114.96	OECD (1985)
Pu_2S_3 (c)	−233.99	OECD (1985)
PuO_2 (c)	−238.53	OECD (1985)
$Pu(OH)_5^{1-}$ (aq)	−378.10	OECD (1985)
PuO_2^{1+} (aq)	−203.10	OECD (1985)
$PuO_2(OH)_2CO_3^{2-}$ (aq)	−413.99	OECD (1985)
$Pu(OH)_4$ (c)	−340.82	OECD (1985)
$Pu_2(CO_3)_3$ (c)	−697.44	OECD (1985)

Abbreviations see Table 1

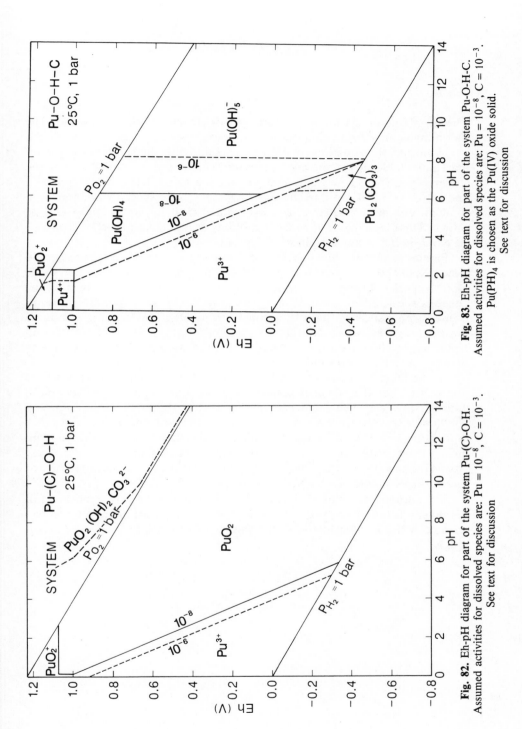

Fig. 83. Eh-pH diagram for part of the system Pu-O-H-C. Assumed activities for dissolved species are: $Pu = 10^{-8}$, $C = 10^{-3}$. $Pu(PH)_4$ is chosen as the Pu(IV) oxide solid. See text for discussion

Fig. 82. Eh-pH diagram for part of the system Pu-(C)-O-H. Assumed activities for dissolved species are: $Pu = 10^{-8}$, $C = 10^{-3}$. See text for discussion

NEPTUNIUM

The Eh-pH diagrams for neptunium species are shown in Figs. 84–86. The thermodynamic data for important neptunium species are given in Table 53.

Similar to the situation for Pu (Figs. 82 and 83), the species NpO_2 is remarkably stable in the presence of water. This is shown in Fig. 84 for the system Np-C-O-H. NpO_2 is dominant over the small fields of Np^{3+}, Np^{4+}, and $Np(OH)_5^-$. NpO_2 will oxidize to NpO_2^+, however, especially under neutral to acidic conditions. There also exists a small field for $NpO_2(OH)_2CO_3^{2-}$ as shown. This diagram attests to the immobility of Np in natural settings provided NpO_2 is the dominant phase of Np(IV).

Figures 85 and 86 show the drastic changes in the Eh-pH diagrams should either $NpO(OH)_2$ (Fig. 85) of $Np(OH)_4$ (Fig. 86) be chosen as the stable Np(IV) species. In Fig. 85 it is noted that the Eh-pH space occupied by the aqueous species, Np^{3+}, Np^{4+}, and especially NpO_2^+ and $Np(OH)_5^-$ have increased with respect to that of the solid, $NpO(OH)_2$. This effect is even more drastic in Fig. 86, where $Np(OH)_4$ is the Np(IV) phase. Here, $Np(OH)_5^-$ and NpO_2^+ are the dominant species, and the field of the $Np(OH)_4$ is small. Fields of Np^{3+} and Np^{4+} also remain small.

At the Oklo Natural Reactor, Np did not, like Pu, migrate. Part of this is due to the fact that there is an almost ideal solid solution between PuO_2 and/or NpO_2 with UO_2, and part due to the fact that apparently PuO_2 and NpO_2 were the preferred M(IV) species where the host pitchblende did break down (Brookins 1987b). Under man-stored radioactive waste repositories, with an unknown water/rock ratio, other Np(IV) phases such as $NpO(OH)_2$ or $Np(OH)_4$ may form, thus increasing potential for Np migration. Both of these species should age to NpO_2, however.

The author (Brookins 1978b,c, 1979b,1984) has previously published Eh-pH diagrams for Np species, but in this work the data used are of a much better quality. Hence, Figs. 84–86 published herein should be referenced accordingly.

Table 53. Thermodynamic data for neptunium

Species (state)	ΔG_f^0 (kcal/gfw)	Reference
Np^{3+} (aq)	-123.59	OECD (1985)
Np^{4+} (aq)	-120.20	OECD (1985)
$Np(OH)_5^{1-}$ (aq)	-384.08	OECD (1985)
NpO_2 (c)	-244.22	OECD (1985)
NpO_2^{1+} (aq)	-218.69	OECD (1985)
NpO_2^{2+} (aq)	-190.20	OECD (1985)
$NpO_2(OH)_2$ (am)	-294.57	OECD (1985)
$Np(CO_3)_2$ (c)	-389.63	OECD (1985)
$NpO_2(OH)_2CO_3^{2-}$ (aq)	-424.05	OECD (1985)
Np_2O_3 (c)	-346.08	OECD (1985)
Np_2O_5 (c)	-481.12	OECD (1985)
$NpO_2(OH)_2$ (c)	-294.57	OECD (1985)

Abbreviations see Table 1

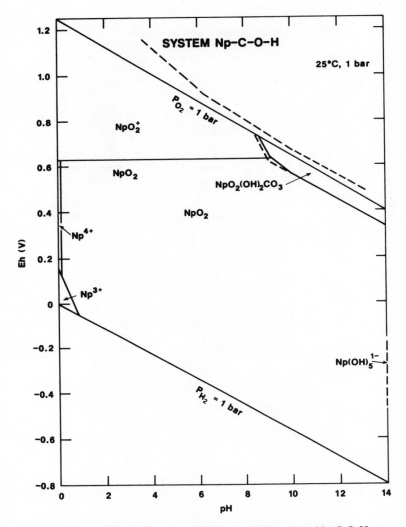

Fig. 84. Eh-pH diagram for part of the system Np-C-O-H.
Assumed activities for dissolved species are: $Np = 10^{-8}$, $C = 10^{-3}$. NpO_2 is chosen as the dominant Np(IV) oxy-solid.
See text for discussion

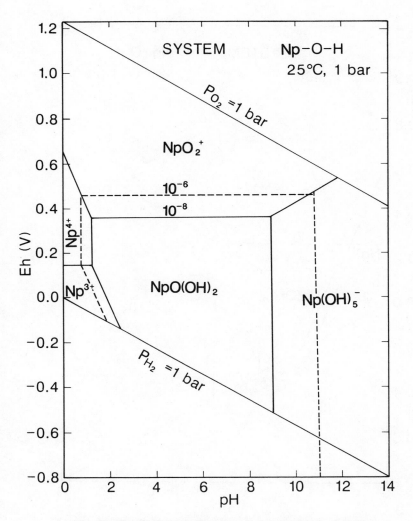

Fig. 85. Eh-pH diagram for part of the system Np-C-O-H.
Assumed activities for dissolved species are: $Np = 10^{-8, -6}$, $C = 10^{-3}$. $NpO(OH)_2$ is chosen as the dominant Np(IV) oxy-solid.
See text for discussion

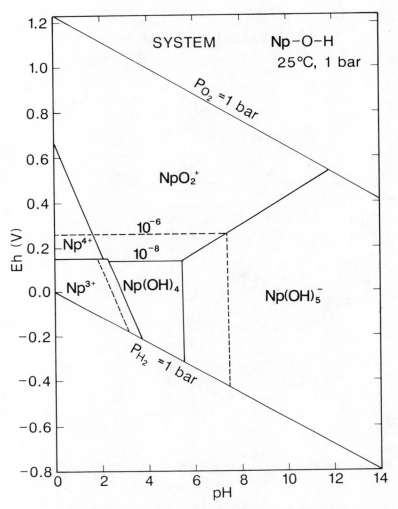

Fig. 86. Eh-pH diagram for part of the system Np-C-O-H.
Assumed activities for dissolved species are: Np = $10^{-6, -8}$, C = 10^{-3}. Np(OH)$_4$ is chosen as the dominant Np(IV) oxy-solid.
See text for discussion

URANIUM

The Eh-pH diagrams for uranium species are shown in Figs. 87–91. The thermodynamic data for important uranium species are given in Table 54.

The most commonly produced Eh-pH diagram for U species is shown in Fig. 87 for the system U-C-O-H with part of the system Fe-S-O-H superimposed. This diagram is extremely useful for depicting phase relationships in these systems with respect to formation of low-grade uranium deposits, and the probable range of Eh-pH for such deposits is indicated in the shaded-edged zone. In Fig. 87 it is assumed that stoichiometric UO_2 is the U(IV) phase, and that dissolved U is transported as uranyl ion or the various eranyl carbonate complexes $[UO_2CO_3^0, UO_2(CO_3)_2^{2-}, UO_2(CO_3)_3^{4-}]$. Brookins (1977, 1979b) has discussed this diagram in detail. It is noteworthy that the activity contours for dissolved U fall above the sulfide-sulfate boundary (marked by $FeS_2:Fe^{2+}$ or Fe_2O_3). This presents a problem as there is abundant evidence that much U ore actually forms at or very near the sulfide-sulfate boundary (Brookins 1977, 1979a, 1984). Lowering the activity of dissolved U to values near and below 10^{-12} implies vanishingly small amounts of U being transported, and this is not satisfactory. In Fig. 88, the system U-C-O-H is shown (without superimposed Fe-S-O-H systematics), and now stoichiometric U_3O_8 is included for pitchblende. Although here pitchblende occupies a small part of the Eh-pH diagram, its field does expand with impurities, higher U(VI):U(IV) ratio, such that at $a_U = 10^{-8}$ the lower part of the pitchblende field will very nearly be coincident with the sulfide-sulfate fence.

Figure 89 shows part of the system U-O-H without any carbonate. What is striking is the large field of U_3O_8 which occurs. This is significant as this may explain the large U deposits of suspected sedimentary origin with cogenetic hematite and pitchblende.

In the presence of Si, the U diagram becomes more interesting. In the system U-Si-C-O-H (Fig. 90), an important field of coffinite (as $USiO_4$) appears. Two contours for different Si activities (10^{-1}, $10^{-3.5}$) are shown. For $a_{Si} = 10^{-1}$, the field of coffinite (Fig. 90, dashed line) persists to pH = 9 before dissolving to $U(OH)_5^-$. By comparison to Fig. 87, the sulfide-sulfate fence would cross this boundary between pH 7 to 8, thus some cogenetic coffinite-pyrite is easily explained. In cases where the activity of Si may decrease as coffinite precipitates, then it will be replaced by pitchblende until the Si activity builds up again. Hence, alternating layers of coffinite and pitchblende will result. This is entirely consistent with many sedimentary U deposits.

Finally, in the absence of dissolved C, coffinite shows a stability field below pH = 7. Hence, coffinite would not be expected in the hematite-pitchblende high-grade sedimentary ores, and this is consistent with observation. Similarly, under acidic reducing conditions and in carbonate-poor environments, coffinite may form.

Numerous Eh-pH diagrams for U species are present in the literature. In addition to the author's previous efforts (Brookins 1976a, 1977, 1978b, 1979a, c, 1982, 1984), there are those presented by Garrels (1959), Garrels and Christ (1965), Langmuir (1978), and Lemire and Tremain (1980). The reader is referred to these sources for more detail. Not covered in Figs. 87–91 are the several U(VI) solid species; these have been discussed by Garrels and Christ (1965), Brookins (1981), and in great depth by Tripathi (1984).

Table 54. Thermodynamic data for uranium

Species (state)	ΔG_f^0 (kcal/gfw)	Reference
UO_2 (c)	−246.62	OECD (1985)
$U(OH)_5^{1-}$ (aq)	−389.77	OECD (1985)
U_3O_8 (c)	−805.35	OECD (1985)
UO_2^{2+} (aq)	−227.66	OECD (1985)
UO_2CO_3 (c)	−373.50	OECD (1985)
UO_2CO_3 (aq)	−367.57	OECD (1985)
$UO_2(CO_3)_2^{2-}$ (aq)	−502.97	OECD (1985)
$UO_2(CO_3)_3^{4-}$ (aq)	−635.57	OECD (1985)
$USiO_4$ (c)	−445.00	Langmuir (1978)

Abbreviations see Table 1

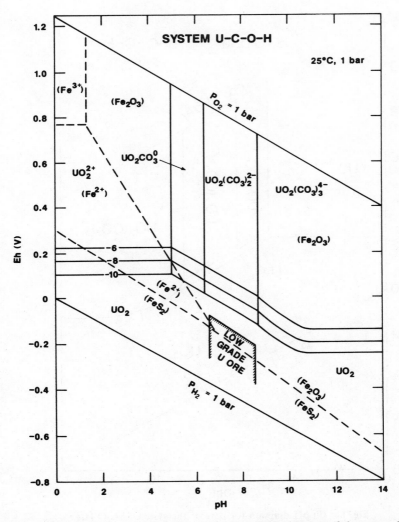

Fig. 87. Eh-pH diagram for part of the system U-C-O-H with part of the system Fe-S-O-H superimposed.
Assumed activities for dissolved species are: $U = 10^{-6, -8, -10}$, $C = 10^{-3}$, $Fe = 10^{-6}$, $S = 10^{-3}$. Area marked *U Ore* from Brookins (1982). See text for details

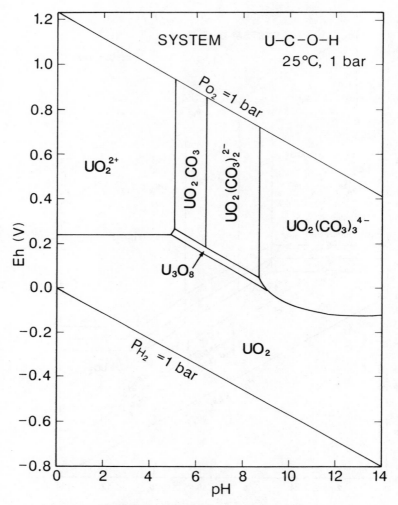

Fig. 88. Eh-pH diagram for part of the system U-C-O-H. Assumed activities for dissolved species are: U = $10^{-8,\ -6}$, C = 10^{-3}. See text for discussion

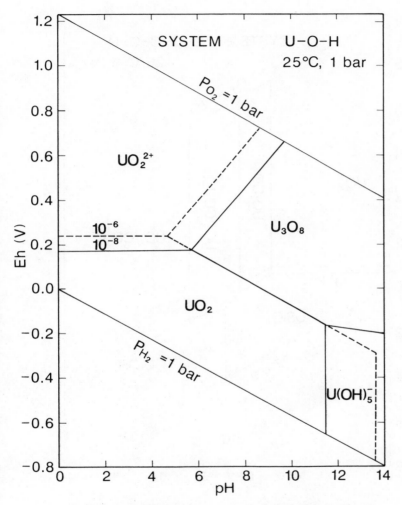

Fig. 89. Eh-pH diagram for part of the system U-O-H. Assumed activity for dissolved U = $10^{-6, -8}$. Note the large field of U_3O_8 in this C-absent diagram. See text for discussion

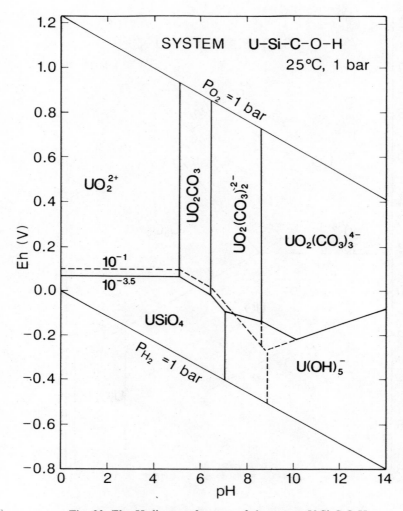

Fig. 90. Eh-pH diagram for part of the system U-Si-C-O-H.
Assumed activities for dissolved species are: $U = 10^{-8}$, $C = 10^{-3}$, $Si = 10^{-3, -1}$.
See text for discussion

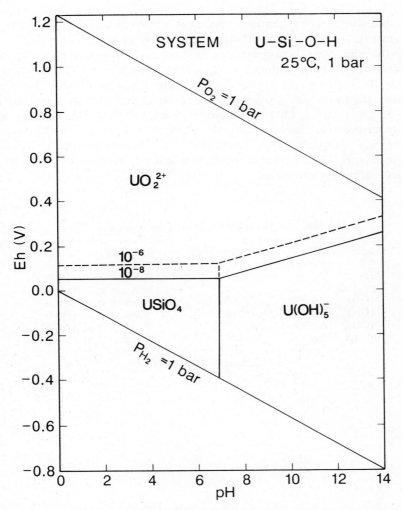

Fig. 91. Eh-pH diagram for part of the system U-Si-O-H.
Assumed activities for dissolved species are: $U = 10^{-8, -6}$, $Si = 10^{-3.5}$.
See text for discussion

THORIUM

The Eh-pH diagram for thorium species is shown in Fig. 92. The thermodynamic data for important thorium species are given in Table 55.

The thorium Eh-pH diagram for the system Th-S-O-H (Fig. 92) is dominated by the large stability field of ThO_2. In nature, $Th(OH)_4$ may precipitate first, but it will age to ThO_2 quickly. The species ThS_2 is stable only below the lower stability limit of water. At higher Eh and below pH 3 ($a_{Th} = 10^{-8}$, $a_S = 10^{-3}$), ThO_2 will dissolve to form $ThSO_4^+$. This species, as well as other possible Th-sulfate complexes, may be important for Th transport from uranium mill tailings (including ^{230}Th formed from ^{238}U decay), but at higher pH Th will be effectively immobilized. Th does not form any important carbonate complexes.

Table 55. Thermodynamic data for thorium

Species (state)	ΔG_f^0 (kcal/gfw)	Reference
ThO_2 (c)	−279.34	Wagman et al. (1982)
Th^{4+} (aq)	−168.52	Wagman et al. (1982)
$Th(OH)^{3+}$ (aq)	−220.00	Wagman et al. (1982)
$Th(OH)_2^{2+}$ (aq)	−272.68	Wagman et al. (1982)
$Th(OH)_6^{2-}$ (aq)	−445.00	Wagman et al. (1982)
$Th(SO_4)^{2+}$ (aq)	−353.90	Wagman et al. (1982)

Abbreviations see Table 1

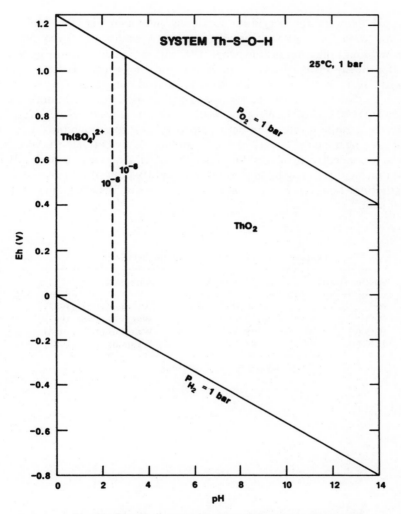

Fig. 92. Eh-pH diagram for part of the system Th-S-O-H. Assumed activities for dissolved species are: Th = $10^{-6, -8}$, S = 10^{-3}. See text for discussion

BERYLLIUM

The Eh-pH diagram for beryllium species is shown in Fig. 93. The thermodynamic data for important beryllium species are given in Table 56.

The beryllium diagram is somewhat difficult to explain. From Fig. 93, it is seen that in the system Be-O-H the important solid species is BeO, which dissolves to form Be^{2+} under acidic to near-neutral conditions (pH = 6.2 or 5.2 for $a_{Be} = 10^{-6}$ and 10^{-4}, respectively) and to BeO_2^{2-} under more basic conditions (pH = 12.5 or 13.5 for $a_{Be} = 10^{-6}$ and 10^{-4}, respectively). Yet, Be is known to be quite insoluble in nature, as beryl is a common constituent in some placers near acidic rocks, and bertrandite, $Be_4Si_2O_7(OH)_2$, is also insoluble and concentrated in residual deposits of silicic volcanic rocks. It is possible that $Be(OH)_2$ may be important in initial fixation of Be released during weathering. $Be(OH)_2$ is metastable with respect to BeO but only by less than 0.6 kcal.

Table 56. Thermodynamic data for beryllium

Species (state)	ΔG_f^0 (kcal/gfw)	Reference
Be^{2+} (aq)	−90.76	Wagman et al. (1982)
BeO_2^{2-} (aq)	−152.99	Wagman et al. (1982)
BeO (c)	−138.70	Wagman et al. (1982)
$Be(OH)_2$ (c)	−194.79	Wagman et al. (1982)

Abbreviations see Table 1

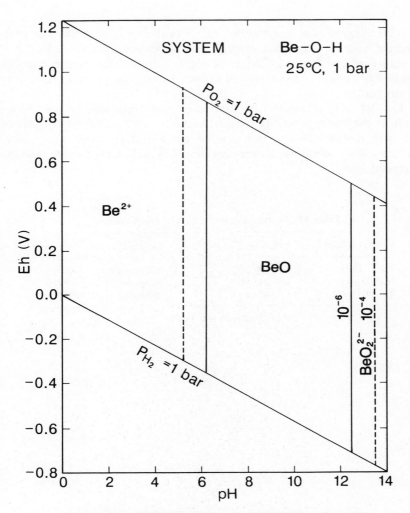

Fig. 93. Eh-pH diagram for part of the system Be-O-H. Assumed activity of dissolved Be = $10^{-6, -4}$. See text for discussion

MAGNESIUM

The Eh-pH diagram for magnesium species is shown in Fig. 94. The thermodynamic data for important magnesium species are given in Table 57.

In nature, magnesium is known to be highly soluble under acidic to mildly basic solutions, and is concentrated in sea water as the second most dominant cation (to Na^+).

In Fig. 94 it is assumed that total dissolved carbonate is low, hence $Mg(OH)_2$ is the stable solid Mg species. For $a_C = 10^{-2}$ of higher, a field of $MgCO_3$ will replace the low pH side of the $Mg(OH)_2$ field. Magnesium sulfates are known in nature; all are water-bearing and highly soluble, and are not included here.

Table 57. Thermodynamic data for magnesium

Species (state)	ΔG_f^0 (kcal/gfw)	Reference
Mg^{2+} (aq)	−108.70	Wagman et al. (1982)
$MgCO_3$ (c)	−241.90	Wagman et al. (1982)
$Mg(OH)_2$ (c)	−199.21	Wagman et al. (1982)

Abbreviations see Table 1

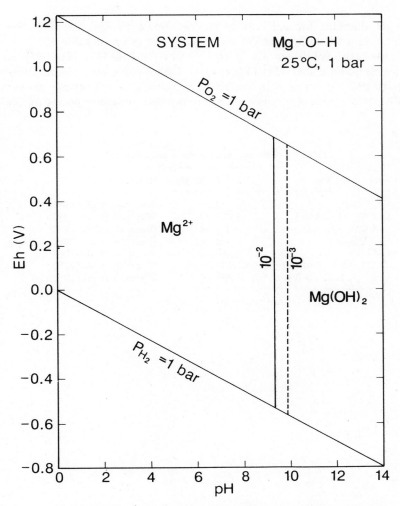

Fig. 94. Eh-pH diagram for part of the system Mg-O-H. Assumed activity of dissolved Mg = $10^{-2, -3}$. See text for discussion

CALCIUM

The Eh-pH diagram for calcium species is shown in Fig. 95. The thermodynamic data for important calcium species are given in Table 58.

The important solids shown in the system Ca-C-O-H-S (Fig. 95) are $CaCO_3$ (calcite) and $CaSO_4 \cdot 2H_2O$ (gypsum). Under some surface conditions, anhydrite ($CaSO_4$) is stable with respect to gypsum, and will replace gypsum on the Eh-pH diagram. Ca^{2+} is the only important aqueous species of Ca.

Table 58. Thermodynamic data for calcium

Species (state)	ΔG_f^0 (kcal/gfw)	Reference
Ca^{2+} (aq)	−132.31	Wagman et al. (1982)
$CaCO_3$ (c) (calcite)	−269.79	Wagman et al. (1982)
$CaCO_3$ (c) (aragonite)	−269.54	Wagman et al. (1982)
$Ca(OH)_2$ (c)	−214.74	Wagman et al. (1982)
CaO (c)	−144.37	Wagman et al. (1982)
$CaSO_4$ (c) (anhydrite)	−315.92	Wagman et al. (1982)
$CaSO_4 \cdot 2H_2O$ (c) (gypsum)	−429.56	Wagman et al. (1982)

Abbreviations see Table 1

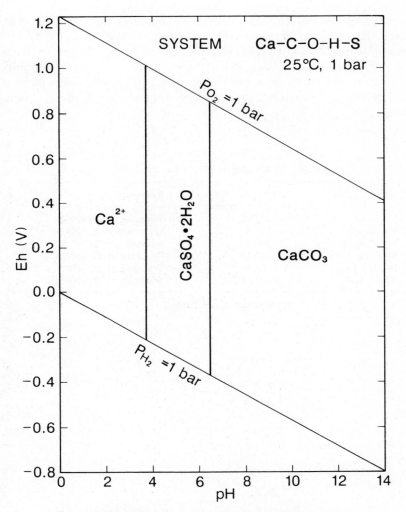

Fig. 95. Eh-pH diagram for part of the system Ca-C-O-H-S. Assumed activities for dissolved species are: Ca = $10^{-2.5}$, S = 10^{-3}, C = 10^{-3}. See text for discussion

STRONTIUM

The Eh-pH diagram for strontium species is shown in Fig. 96. The thermodynamic data for important strontium species are given in Table 59.

In the system Sr-C-S-O-H (Fig. 96), the insolubility of Sr species from mildly acidic to increasingly basic pH is shown. Under very basic to mildly basic conditions, $SrCO_3$ forms a wide range of Eh-pH space, followed by a large field of $SrSO_4$ at intermediate to mildly acidic pH. Sr^{2+} is the dominant aqueous species.

Table 59. Thermodynamic data for strontium

Species (state)	ΔG_f^0 (kcal/gfw)	Reference
Sr^{2+} (aq)	−133.72	Wagman et al. (1982)
$SrCO_3$ (c)	−272.49	Wagman et al. (1982)
$Sr(OH)_2$ (c)	−207.80	Latimer (1952)
SrO (c)	−134.30	Wagman et al. (1982)
$SrOH^+$ (aq)	−172.39	Wagman et al. (1982)
$SrSO_4$ (c)	−320.48	Wagman et al. (1982)

Abbreviations see Table 1

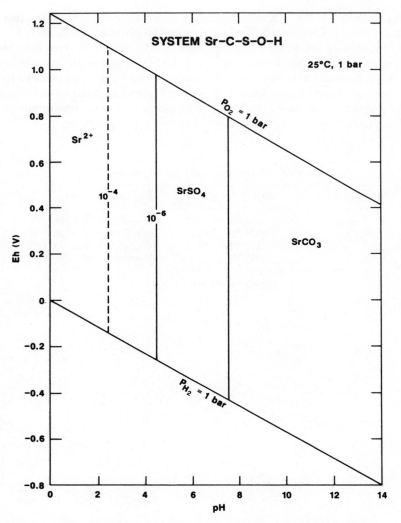

Fig. 96. Eh-pH diagram for part of the system Sr-C-S-O-H. Assumed activities for dissolved species are: $Sr = 10^{-6,\ -4}$, $S = 10^{-3}$, $C = 10^{-3}$. See text for discussion

BARIUM

The Eh-pH diagram for barium species is shown in Fig. 97. The thermodynamic data for important barium species are given in Table 60.

The Eh-pH diagram (Fig. 97) is very similar to that for Ra (see Fig. 98) in the system Ba-O-H-C-S. Below pH of 1.2, $BaSO_4$ dissolves to form Ba^{2+}, and $BaSO_4$ is replaced by $BaCO_3$ at about pH = 11.6. Thus, over most of the Eh-pH field of water, Ba is immobile as $BaSO_4$ and is fixed in soils and rocks. This is fortunate for Ba is a highly toxic element used in large quantities for drilling mud additives and for medicinal and other purposes.

Table 60. Thermodynamic data for barium

Species (state)	ΔG_f^0 (kcal/gfw)	Reference
Ba^{2+} (aq)	−134.03	Wagman et al. (1982)
$BaSO_4$ (c)	−325.57	Wagman et al. (1982)
$BaCO_3$ (c)	−271.89	Wagman et al. (1982)
BaO (c)	−125.50	Wagman et al. (1982)

Abbreviations see Table 1

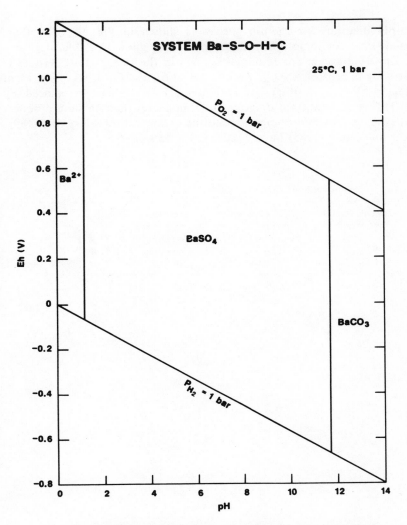

Fig. 97. Eh-pH diagram for part of the system Ba-S-O-H-C. Assumed activities for dissolved species are: Ba = 10^{-6}, C = 10^{-3}, S = 10^{-3}. See text for discussion

RADIUM

The Eh-pH diagram for radium species is shown in Fig. 98. The thermodynamic data for important radium species are given in Table 61.

The Eh-pH diagram for radium (Fig. 98) in the system Ra-O-H-C-S is dominated by a field of $RaSO_4$. At low pH (2.6), $RaSO_4$ dissolves to form Ra^{2+} ($a_{Ra} = 10^{-8}$, $a_S = 10^{-3}$). At high pH (12.4), $RaSO_4$ is replaced by $RaCO_3$. This is an important diagram for it indicates that Ra should be immobile in the natural environment due to the extreme insolubility of $RaSO_4$. Langmuir and Riese (1985) have discussed Ra data in detail.

Table 61. Thermodynamic data for radium

Species (state)	ΔG_f^0 (kcal/gfw)	Reference
Ra^{2+} (aq)	−134.20	Wagman et al. (1982)
$RaSO_4$ (c)	−326.39	Wagman et al. (1982)
$RaCO_3$ (c)	−271.69	Langmuir and Riese (1985)

Abbreviations see Table 1

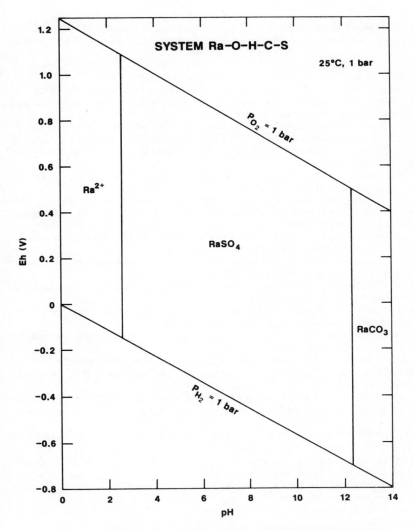

Fig. 98. Eh-pH diagram for part of the system Ra-O-H-C-S. Assumed activities for dissolved species are: Ra = 10^{-8}, S = 10^{-3}, C = 10^{-3}. See text for discussion

References

Anderson JA (1982) Characteristics of leached capping and techniques of appraisal. In: Titley SR (ed) Advances of the porphyry copper deposits. Univ Arizona Press, Tucson, p 275–297

Baas Becking LGM, Kaplan IR, Moore D (1960) Limits of the natural environment in terms of pH and oxidation-reduction potential. J Geol 68:243–284

Baes CF, Mesmer RE (1976) The hydrolysis of cations. Wiley Inc., New York, 489 p

Barner HE, Scheuerman RV (1978) Handbook of thermochemical data for compounds and aqueous species. Wiley & Sons, New York, 156 p

Berner RA (1971) Principles of chemical sedimentology. McGraw-Hill, New York, 240 p

Bird GW, Lopata VJ (1980) Solution interaction of nuclear waste anions with selected geologic materials. In: Northrup CJ (ed) Sci Basis Noc Wste Mngt II. Plenum Press, New York, pp 419–427

Brookins DG (1976a) Uranium deposits of the Grants, New Mexico mineral belt. US Energy Res Dev Agency Rpt, GJO-1636-1, 120 p

Brookins DG (1976b) The Grants mineral belt, New Mexico. Comments on the coffinite-uraninite relationship, probable clay mineral reactions, and pyrite formation. New Mexico Geol Soc Spec Pub 6:255–269

Brookins DG (1977) Uranium deposits of the Grants mineral belt. Geochemical constraints. Rocky Mt Assoc Geol Gdbk, pp 337–352

Brookins DG (1978a) Retention of transuranic, other actinide elements and bismuth at the Oklo natural reactor, Gabon. Chem Geol 23:307–323

Brookins DG (1978b) Eh-pH diagrams for elements from $Z = 40$ to $Z = 52$. Application to the Oklo natural reactor. Chem Geol 23:324–341

Brookins DG (1978c) Application of Eh-pH diagrams to problems of retention and/or migration of fissiogenic elements at Oklo. IAEA, pp 243–266

Brookins DG (1979a) Uranium deposits of the Grants, New Mexico mineral belt II. U.S.D.O.E. Rpt, BFEC-GJO-76-029E, 411 p

Brookins DG (1979b) Thermodynamic considerations underlying the migration of radionuclides in geomedia. Oklo and other examples. In: McCarthy GJ (ed) Sci Basis Nuc Wste Mngt I. Plenum Press, New York, pp 355–366

Brookins DG (1979c) Eh-pH diagrams for elements of interest at the Oklo natural reactor at 25°C, 1 bar pressure and 200°C, 1 bar pressure. Los Alamos NH Lab, CNC-11 (unpubl.)

Brookins DG (1981) Primary uranophane from the Ambrosia Lake uranium district, Grants mineral belt, USA. Mineral Deposita 16:3–7

Brookins DG (1982) Geochemistry of clay minerals for uranium exploration in the Grants mineral belt, New Mexico. Mineral Deposita 17:37–53

Brookins DG (1983a) Eh-pH diagrams for the rare earth elements at 25°C and one bar pressure. Geochem J 17:223–229

Brookins DG (1983b) Migration and retention of elements at the Oklo natural reactor. Environ Geol 4:201–208

Brookins DG (1984) Geochemical aspects of radioactive waste disposal. Springer, Berlin Heidelberg New York Tokyo, 347 p

Brookins DG (1986a) Geochemical behavior of antimony, arsenic, cadmium and thallium. Eh-pH diagrams for 25°C, 1 bar pressure. Chem Geol 54:271–278

Brookins DG (1986b) Rhenium as analog for fissiogenic technetium. Eh-pH diagram (25°C, 1 bar) constraints. J Appld Geochem 1:513–517

Brookins DG (1987a) Platinoid element Eh-pH diagrams (25°C, 1 bar) in the systems M-O-H-S with geochemical applications. Chem Geol 64:1–8

Brookins DG (1987b) Radionuclide behavior at the Oklo natural reactor. Oak Ridge National Lab, Tech Manuscript No 10151, 254 p

Brookins DG, Thomson BM, Longmire PA (1982) Early diagenesis of uranium mine stope backfill. Proc Fifth Ann Uranium Mill Tailings Mngt Conf, Fort Collins, Colorado, pp 27–37

Cabri LJ (ed) (1981) Platinum-group elements. Mineralogy, geochemistry, recovery. Canad Inst Min Metal Spec 23:265

Curtis DB, Benjamin TM, Gancarz AJ (1983) The Oklo reactors. Natural analogs to nuclear waste repositories. Los Alamos Ntl Lab Rep LA-UR-81 3:32

Dove PM, Rimstidt JD (1985) The solubility and stability of scorodite, $FeASO_4 \cdot 2H_2O$. Am Mineral 70:838–844

Drever JI (1982) The geochemistry of natural waters. Prentice Hall, New York, 388 p

Elderfield H, Greaves MJ (1982) The rare earth elements in sea water. Nature (Lond) 296:214–219

Garrels RM (1959) Mineral equilibria. Addison-Wesley Reading Mass, 349 p

Garrels RM, Christ CL (1965) Minerals, solutions and equilibria. Harper and Rowley, New York, 453 p

Gerasimov YaI, Krestovnikov AN, Shakov AS (1963) Metallurgy of the rare metals. Metallurgizdat, Leningrad (in Russian)

Goldberg ED, Hodge V, Kay P, Stallard M, Koide M (1986) Some comparative marine chemistries of platinum and iridium. J Appl Geochem 1:227–232

Hem JD (1977) Reactions of metal ions at surfaces of hydrous iron oxides. Geochim Cosmochim Acta 41:527–538

Hem JD (1981) Redox processes at surfaces of manganese oxide and their effects on aqueous metal ions. Chem Geol 21:199–218

Henderson P (1984) Rare earth element geochemistry. Elsevier, Amsterdam, 510 p

Hostetler JD (1984) Electrodes, electrons, aqueous electrons, and redox potentials in natural waters. Am J Sci 284:731–746

Howard JJ (1977) Geochemistry of selenium. Formation of ferroselite and selenium behavior in the vicinity of oxidizing sulfide and uranium deposits. Geochim Cosmochim Acta 41:1665–1678

Kahlil MY, White WB (1984) Dissolution of technetium from nuclear waste forms. In: McVay GL (ed) Sci Basis Nuc Wste Mngt VII. Elsevier, New York, pp 655–666

Klein C, Bricker OP (1977) Some aspects of the sedimentary and diagenetic environment of Proterozoic banded iron formation. Econ Geology 72:1457–1470

Krauskopf KB (1979) Introduction to geochemistry, 2nd edn. McGraw-Hill, New York, 617 p

Krauskopf KB (1986) Thorium and rare-earth metals as analogs for actinide elements. Chem Geol 55:323–335

Langmuir D (1978) Uranium solution-mineral equilibria at low temperatures with applications to sedimentary ore deposits. Geochim Cosmochim Acta 42:547–570

Langmuir D, Riese AC (1985) The thermodynamic properties of radium. Geochim Cosmochim Acta 49:1593–1602

Latimer WM (1952) Oxidation potentials, 2nd edn, 5th printing. Prentice Hall, Princeton, NJ, 392 p

Lemire RJ, Tremaine PR (1980) Uranium and plutonium equilibria in aqueous solutions to 200 °C. J Chem Engg Data 25:361–370

Lindberg JW, Runnells DD (1984) Ground water redox reactions. An analysis of equilibrium state applied to Eh measurements and geochemical modeling. Science 225:925–927

Lindsay WL (1979) Chemical equilibria in soils. Wiley & Sons, New York, 449 p

Longmire PA, Hicks RT, Brookins DG (1981) Geochemical interactions between ground water and uranium mine stope backfilling – Grants mineral belt, New Mexico. Application of Eh-pH diagrams. Proc Fourth Ann Uranium Mill Tailings Mngt Conf, Fort Collins, Colorado, pp 389–414

Maynard JB (1983) Geochemistry of sedimentary ore deposits. Springer, Berlin Heidelberg New York, 305 p

Nordström DK, Jenne EA, Ball JW (1979) Redox equilibria of iron in acid mine waters. In: Jenne EA (ed) Chemical modeling of aqueous systems. Am Chem Soc Sym Ser 33:51–80

OECD (1985) Organization for economic co-operation and development. Compilation of selected thermodynamic data (provided by Muller AB, Paris)

Parks GA, Nordström DK (1979) Estimated free energies of formation, water solubilities, and stability fields for schutteite ($Hg_3O_2SO_4$) and corderoite ($Hg_3S_2Cl_2$) at 298 K. In: Jenne EA (ed) Chemical modelling in aqueous systems. Amer Chem Soc Sym Series 93, Washington, DC, pp 339–352

Piper DZ (1974) Rare earth elements in the sedimentary cycle, A review. Chem Geol 14:285–301

Pourbaix M (1966) Atlas of electrochemical equilibria. Pergamon Press, Oxford, 645 p

Rard JA (1983) Critical review of the chemistry and thermodynamics of technetium and some of its inorganic compounds and aqueous species. Lawrence Livermore Ntl Lab Rpt, LLNL UCRL-53440, 87 p

Robie RA, Hemingway BS, Fisher JR (1978) Thermodynamic properties of minerals and related substances at 290.15 K and one bar (10^5 pascals) pressure and at higher temperatures. US Geol Surv Bull 1452:456

Schumm RH, Wagman DD, Bailey S, Evans WH, Parker VB (1973) Selected values of chemical thermodynamic properties. Tables for the lanthanide (rare earth) elements. Ntl Bur Standards Tech Note 270–7:75

Smith RM, Martell AE (1976) Critical stability constants, Vol 4. Inorganic complexes. Plenum Press, New York, 176 p

Stumm W, Morgan JJ (1981) Aquatic chemistry, 2nd edn. Wiley Inc, New York, 780 p

Thomson BM, Longmire PA, Brookins DG (1986) Geochemical constraints on underground disposal of uranium mill tailings. J Appl Geochem 1:335–344

Titley SR, Anthony JW (1961) Some preliminary observations on the theoretical geochemistry of molybdenum under supergene conditions. Ariz Geol Soc Digest 4:103–116

Tripathi VS (1984) Uranium (VI) transport modeling. Geochemical data and submodels. Ph. D. Dissertation, Stanford University, 345 p (unpubl.)

Verink ED (1979) Simplified procedure for constructing Pourbaix diagrams. J Educ Modules Mat Sci Engg 1:535–560

Wagman DD, Evans WH, Parker VB, Halow I, Bailey SM, Schumm RH (1969) Selected values of chemical thermodynamic properties. Tables for the elements 35 through 53 in the standard order of arrangement. Ntl Bur Stand Tech Note 270–4:141

Wagman DD, Evans WH, Parker VB, Schumm RH, Halow I, Bailey SM, Churney KL, Buttall RL (1982) The NBS tables of chemical thermodynamic properties. Selected values for inorganic and C1 and C2 organic substances in SI units. J Phys Chem Ref Data 11, suppl 2:392

Westland AD (1981) Inorganic chemistry of the platinum-group elements. In: Cabri, pp 5–18

Winters GV, Buckley DE (1986) The influence of dissolved $FeSi_3O_3(OH)_8^{\circ}$ on chemical equilibria in pore waters from deep sea sediments. Geochim Cosmochim Acta 50:277–288

Zoubov N de, Pourbaix M (1966) Rhenium. In: Pourbaix M (ed), pp 300–306

Subject Index

Abbreviations for tables 13
Actinium 13
Alabandite 94
Aluminum 44, 45
Americium 13, 141–143
Anatase 114
Angelsite 42
Antimony 30, 31
Argentite 64
Astatine 13
Azurite 60

Baddeleyik 116
Barium 168, 169
Bertrandite 160
Beryllium 160, 161
Bismuth 32, 33
Bismuthinite 32
Boron 44, 45
Boronite 60
Bromine 13, 14
Brookite 114

Cadmium 56, 57
Calcium 164, 165
Calomel 58
Cerium 122–125, 139
Cerussite 42
Cesium 13
Chalcocite 60
Chalcoppyrite 60
Chlorine 13, 14
Chlorite 108
Chromium 102, 103
Claudetite 28
Cobalt 71, 72
Coffinite 104, 151, 152
Cooperite 88
Copper 60–63
Corellite 60

Cuprite 60
Curium 13

Dysposium 122, 123, 132

Eh-pH: general 1–13
Erbium 122, 123, 130
Erlichmanite 92
Europium 122–124, 135

Ferroselite 18
Fluorine 13, 14
Francium 13

Gadolinium 122, 123, 134
Gallium 48, 49
Germanium 38, 39
Goethite 73
Gold 66, 67
Greenockite 54
Gypsum 162

Hafnium 118, 119
Haggite 108
Helium 12
Hematite 73
Holmium 122, 123, 131
Hydrogen 12, 13

Indium 50, 51
Iodine 14, 15
Iridium 90, 91
Iron 73–81

Jordisite 104

Krypton 12

Lanthanum 122–124, 140
Laurite 86

175

Lead 42, 43
Litharge 42
Lithium 13
Lutetium 122, 123, 127

Magnesium 162, 163
Malachite 60
Manganese 94−96
Massicot 42
Mercury 58, 59
Methane 34
Millerite 68
Minium 42
Molybdenite 104
Molybdenum 104, 105
Montroseite 108
Montroydite 58

Neodymium 122−124, 137
Neon 12
Neptunium 13, 146−150
Nickel 68−70
Niobium 110, 111
Nitrogen 24, 25

Oklo 54, 110, 116, 125, 144, 146
Organic C-carbonate boundary 12
Osmium 92, 93
Oxygen 13

Palladium 82, 83
pE 2
Phosphorus 26, 27
Pitchblende 151
Platinum 88, 89
Plattnerite 42

Plutonium 13, 144, 145
Polonium 13, 22, 23
Potassium 13
Praesodymium 122−124, 138
Promethium 13, 122
Protoactinium 13

Radium 170, 171
Radon 12
Rare earths 122−126
Rhenium 100, 101
Rhodium 84, 85
Rhodochrosite 94
Rubidium 13
Ruthenium 86, 87

Samarium 122−124, 136
Scandium 120, 121
Scorodite 28
Selenium 18, 19
Siderite 73
Silicon 36, 37
Silver 64, 65
Smithsonite 54
Sodium 13
Sphalerite 54
Spherocobaltite 71
Stibiconite 30
Stimnite 30
Strontium 166, 167
Sulfide-sulfate boundary 10, 12
Sulfur 10, 12, 16, 17

Tantalum 112, 113
Technetium 13, 97−99